Applications of Linear Algebra

Applications of Linear Algebra

USIAHON E. JAMES

APPLICATIONS OF LINEAR ALGEBRA

ISBN 10: 1514817306

Printed by Amazon CreateSpace, An Amazon.com company.

www.amazoncreatespace.com/5574593

Author Contact:

E-mail: usiahonjames@yahoo.com

Mobile: +2348074732357

Contents

Preface

Linear algebra finds a natural place in the mathematical modeling of realities; thus, its benefits to physical, biological and social sciences cannot be gainsaid. Moreover, the application of linear algebra concepts to suitable systems helps to achieve deeper insights into such systems and to make useful generalizations. This book, *applications of linear algebra*, has been written to satisfy the requirement of undergraduate students of mathematics, engineering, physics, actuarial science, statistics, chemistry, economics, and as many as need to apply linear algebra to their studies and research. It is also a very useful tool for students preparing for the *SAT2 Mathematics, A- level Cambridge* Mathematics and Further Mathematics, *and GRE subject test* in Mathematics.

The book deals with the application of linear algebra to: linear systems of algebraic equations, least-square approximation, conics and quadratic surfaces, Optics, and systems of constant-coefficient linear ordinary differential equations, as well as intermediary topics needed to make the book a coherent whole. In an attempt to make the subject more substantial and less formal, I have provided the reader with enough computational examples, considering the wide range of my intended audience. However, I have also interspersed this material with some rigorous proofs in order to meet the demand of the more advanced reader as abstraction and rigor cannot be totally eliminated from any useful linear algebra textbook.

Usiahon E. James

Applications of Linear Algebra

Chapter One

Systems of Linear Equations

I assume that the reader has some level of familiarity with the rudiments of linear systems of algebraic equations-at least using the basic solution methods. In this section, however, we will explore the formal ways of dealing with such equations.

SYSTEMS OF LINEAR EQUATIONS

This is simply a collection of linear equations in two or more unknowns. At times the equations are not linear, but can be made linear by a simple change of variables or coordinates transformations. Let's consider an example.

$$a_{11}x_1 + a_{12}x_2 = b_1$$
$$a_{21}x_1 + a_{22}x_2 = b_2$$

This is an example of a system of linear equations with two unknowns, where the constants a_{ij} , b_i ; $1 \leq i, j \leq 2$, are real numbers. This system can be formally solved by first representing it in matrix form; thereafter, we can wield the tools of linear algebra on it to determine its solution(s) if any. The above linear system can be represented as:

$$AX = B \qquad\qquad --(1)$$

$$A = \begin{pmatrix} a_{11} & a_{12} \\ a_{21} & a_{22} \end{pmatrix} , \quad X = \begin{pmatrix} x_1 \\ x_2 \end{pmatrix} , \quad B = \begin{pmatrix} b_1 \\ b_2 \end{pmatrix}$$

The matrix A is called the coefficient matrix, and the vectors X and B are respectively the variable vector and the constant vector. A solution to the above linear system is a vector X_s that satisfies the equation. That is, X_s is such that

$$AX_s = B, \qquad X_s = \begin{pmatrix} x_{1,s} \\ x_{2,s} \end{pmatrix}$$

This (X_s) is called the *Solution vector*. There are several ways of finding X_s. One way is by replacing the system with a simpler but equivalent one (*Gaussian elimination*), such that the variables can be directly and easily determined. Another way is by premultiplying both sides of (1) above by the *inverse matrix* of A; we can also solve the system by using the *Crammer's rule*, the method that involves the evaluation of matrix determinants. Before we look at these different solution methods and even more, let us review some basic elements of matrix terminology and matrix algebra.

MATRICES

A matrix is a rectangular array of elements. The elements are arranged into rows and columns. A matrix with m-number of rows and n-number of columns is referred to as an mxn (pronounced "m by n") matrix. For example,

$$A = \begin{pmatrix} a & b \\ c & d \end{pmatrix} \text{ is a 2 x 2 matrix}$$

$$B = \begin{pmatrix} a & b & c \\ d & e & f \end{pmatrix} \text{ is a 2 x 3 matrix}$$

Also, vectors can be regarded as a row or a column matrix. For instance, the vectors $\begin{pmatrix} a \\ b \\ c \end{pmatrix}$ $and (a \ b \ c)$ are respectively called a column matrix, and a row matrix.

A column matrix, also, can simply be called a column vector, while a row matrix can also be called a row vector. If a matrix has m rows and n columns, the matrix is said to be of order mxn. Note that we can also use square brackets to encase the elements of a matrix instead of the usual brackets used above. We shall make use of both of them interchangeably in later sections.

SCALAR MULTIPLICATION

The scalar multiplication of a matrix A by a real number k, denoted as kA, is defined as the multiplication of each element of A by k. That is, if

$$A = \begin{pmatrix} a & b & c \\ d & e & f \end{pmatrix}, \text{ then}$$

$$kA = \begin{pmatrix} ka & kb & kc \\ kd & ke & kf \end{pmatrix}, k \text{ is real.}$$

ADDITION AND SUBTRACTION OF MATRICES

Not any two matrices are compatible for addition. Two matrices A and B are addition compatible if and only if they both have the same number of rows and columns. For instance, let

$$A = \begin{pmatrix} a & b & c \\ d & e & f \\ g & h & k \end{pmatrix}, \quad B = \begin{pmatrix} l & p & q \\ r & s & t \end{pmatrix}, \quad C = \begin{pmatrix} u & v & w \\ x & y & z \end{pmatrix}$$

Matrices B and C are compatible for addition, while A and B, and A and C are not.

$$B + C = \begin{pmatrix} l+u & p+v & q+w \\ r+x & s+y & t+z \end{pmatrix}$$

that is, two matrices are added by simply adding their corresponding elements.

The Matrix $(B - C)$ is defined as $(B + (-C))$, where $(-C)$ is a scalar multiplication of C by -1. Only matrices of the same order are compatible for addition and subtraction.

MATRIX MULTIPLICATION

Just like matrix addition, not any two matrices are compatible for multiplication. For matrices $A(mxn)$ and $B(pxq)$ to be multipliable, the column number of the "first matrix" must be equal to the row number of the "second "second matrix"". The terms "first matrix" and "second matrix" come in because matrix multiplication, unlike number multiplication, is not *commutative;* that is, in general, $AB \neq BA$. Therefore, in the matrix product AB, A is referred to as the first matrix, and B, the second matrix. Thus, for the product AB to exist, A and B must be compatible for multiplication.

Since A is an m x n matrix, and B is a p x q matrix, multiplication compatibility, in this case, means $n = p$. Therefore, while AB may exist, BA might not. For BA to exist, it is required that $q = m$.

The product matrix $A(mxn)B(pxq)$ is always an m x q matrix, while $B(pxq)A(mxn)$ is a pxn matrix. For example, let

$$A = (a_1, a_2, a_3), \quad B = \begin{pmatrix} b_1 \\ b_2 \\ b_3 \end{pmatrix}, \text{then}$$

$$AB = (a_1, a_2, a_3) \begin{pmatrix} b_1 \\ b_2 \\ b_3 \end{pmatrix} = a_1 b_1 + a_2 b_2 + a_3 b_3$$

Also, if

$$A = \begin{pmatrix} a_{11} & a_{12} \\ a_{21} & a_{22} \end{pmatrix}, \quad B = \begin{pmatrix} a & b & c \\ d & e & f \end{pmatrix}, \text{then}$$

$$AB = \begin{pmatrix} a_{11}a + a_{12}d & a_{11}b + a_{12}e & a_{11}c + a_{12}f \\ a_{21}a + a_{22}d & a_{21}b + a_{22}e & a_{21}c + a_{22}f \end{pmatrix},$$

while BA does not exist.

Matrix – matrix division is not defined.

BASIC TERMINOLOGIES

Square Matrix: A matrix is said to be a square matrix if it has equal number of rows and columns; that is, a square matrix is an n x n matrix.

Identity Matrix: This is a matrix that has all its main-diagonal elements as unity (one) and every other element zero. The main diagonal of a matrix is an imaginary line segment that connects the top left and the bottom right corners of an imaginary square in which the elements of the square matrix are evenly placed. Only square matrices have main diagonal. For example,

$$I_2 = \begin{pmatrix} 1 & 0 \\ 0 & 1 \end{pmatrix}, and\ I_3 = \begin{pmatrix} 1 & 0 & 0 \\ 0 & 1 & 0 \\ 0 & 0 & 1 \end{pmatrix}$$ are identity matrices.

Identity matrices have a special property, such that leaves any matrix unaltered upon multiplication by it. For instance if A and B are any 2x2, and 3x3 matrices respectively, then $AI_2 = I_2A = A$, and $BI_3 = I_3B = B$. They are generally represented as I. An $n x n$ identitiy matrix can be denoted as I_n.

Diagonal Matrix: This is a square matrix in which all the non-main diagonal elements are zero and at least one main diagonal element is not zero. For example, the matrix

$$D = \begin{pmatrix} d_1 & 0 & 0 \\ 0 & d_2 & 0 \\ 0 & 0 & d_3 \end{pmatrix}$$, where, at least, one of $d_1, d_2, d_3 \neq 0$, is a diagonal matrix.

Matrix D can simply be represented as $D = diag(\ d_1, d_2, d_3)$. Therefore, the identity matrices can be viewed as special diagonal matrices in which the diagonal elements are each unity.

Upper Triangular Matrix: This is a square matrix in which the elements below the main diagonal are zero. The following are an example of upper triangular matrices.

$$A = \begin{pmatrix} a_{11} & a_{12} \\ 0 & a_{22} \end{pmatrix}, B = \begin{pmatrix} a_{11} & a_{12} & a_{21} \\ 0 & a_{22} & a_{23} \\ 0 & 0 & a_{33} \end{pmatrix}$$

$$C = \begin{pmatrix} a_{11} & a_{12} & a_{13} & a_{14} \\ 0 & a_{22} & a_{23} & a_{24} \\ 0 & 0 & a_{33} & a_{34} \\ 0 & 0 & 0 & a_{44} \end{pmatrix}$$

If we generally represent the elements of a matrix as a_{ij}, as shown in these examples, where i and j denote, respectively, the row number and the column number of element a_{ij}, then an upper triangular matrix can be defined as that in which $a_{ij} = 0$ for every $i > j$.

Lower Triangular Matrix: Referring to the defined notation a_{ij}, a lower triangular matrix is that in which $a_{ij} = 0$ for every $i < j$.

Transposition of a Matrix: The transposition, or simply transpose, of a matrix A, denoted as A^T, is defined as a matrix in which the rows become the columns and vice versa.

For example, if matrix $A = \begin{pmatrix} a_{11} & a_{12} \\ a_{21} & a_{22} \end{pmatrix}$, then $A^T = \begin{pmatrix} a_{11} & a_{21} \\ a_{12} & a_{22} \end{pmatrix}$; again,

If $A = \begin{pmatrix} a & b & c \\ d & e & f \end{pmatrix}$, then $A^T = \begin{pmatrix} a & d \\ b & e \\ c & f \end{pmatrix}$.

Symmetric Matrix: A matrix is said to be symmetric if it is equal to its transpose; that is, matrix B is symmetric if and only if $B = B^T$. By equality of matrices, we mean matrices whose corresponding elements are equal. For instance, if

$$\begin{pmatrix} a_{11} & a_{12} & a_{13} \\ a_{21} & a_{22} & a_{23} \end{pmatrix} = \begin{pmatrix} b_{11} & b_{12} & b_{13} \\ b_{21} & b_{22} & b_{23} \end{pmatrix}, \text{ then}$$

$$a_{11} = b_{11}, a_{12} = b_{12}, \dots \dots a_{23} = b_{23}.$$

Only matrices of the same order can be equal.

Skew Symmetric: A matrix B is skew symmetric if and only if

$$B^T = -B.$$

Inverse Matrix: The inverse of a matrix A, denoted as A^{-1}, is defined as that matrix that multiples A to give the identity matrix. That is, if B is the inverse of A, then

$$AB = BA = I; \ B = A^{-1}, A = B^{-1}.$$
$$AA^{-1} = A^{-1}A = I.$$

Orthogonal Matrix: A matrix B is said to be orthogonal if and only if

$$B^{-1} = B^T, \ \text{provided that } B^{-1} \text{ exists.}$$

Null Matrix: This is a matrix whose elements are all zero.

Singleton Matrix: This is a 1x1 matrix. It contains only one element; for example, $A = (a)$. Note that $(a) \neq a$, the first is a matrix, the second is a number.

Determinant of a matrix

If $A = \begin{pmatrix} a_{11} & a_{12} \\ a_{21} & a_{22} \end{pmatrix}$, then the determinant of A, denoted as $\det(A)$ or $|A|$, is

$$|A| = \begin{vmatrix} a_{11} & a_{12} \\ a_{21} & a_{22} \end{vmatrix} = a_{11}a_{22} - a_{12}a_{21}$$

Generally, if A is an nxn ($n \geq 2$) matrix, then $|A|$ is obtained by the method of **Laplace expansion** along any row or column. For example, if

$A = \begin{pmatrix} a_{11} & a_{12} & a_{13} \\ a_{21} & a_{22} & a_{23} \\ a_{31} & a_{32} & a_{33} \end{pmatrix}$, then we can find $|A|$ by expanding along the first row, say.

Expanding along the first row means multiplying each of $(-1)^{1+j}a_{1j}$ by its corresponding *minor*, and then adding up. The minor of any element is the determinant of the square submatrix obtained by crossing out the element with two intersecting vertical and horizontal lines. For example, the minors of the elements a_{21} and a_{33}, respectively, are:

$$minor(a_{21}) = \begin{pmatrix} a_{11} & a_{12} & a_{13} \\ a_{21} & a_{22} & a_{23} \\ a_{31} & a_{32} & a_{33} \end{pmatrix} = \begin{vmatrix} a_{12} & a_{13} \\ a_{32} & a_{33} \end{vmatrix} = a_{12}a_{33} - a_{32}a_{13}$$

$$minor(a_{33}) = \begin{pmatrix} a_{11} & a_{12} & a_{13} \\ a_{21} & a_{22} & a_{23} \\ a_{31} & a_{32} & a_{33} \end{pmatrix} = \begin{vmatrix} a_{11} & a_{12} \\ a_{21} & a_{22} \end{vmatrix} = a_{11}a_{22} - a_{12}a_{21}$$

The cofactor of an element a_{ij} of a matrix is defined as $(-1)^{i+j} minor(a_{ij})$

That is,

$$cofactor(a_{ij}) = (-1)^{i+j} minor(a_{ij})$$

The determinant of A, $|A|$, along any row or column is defined as the sum of the products of the elements along the row or column and their corresponding cofactors. That is, if A is an nxn matrix, then,

$$|A| = \sum_{j=1}^{n} a_{1j} \left[(-1)^{1+j} minor(a_{1j}) \right], \text{along row one}$$

$$|A| = \sum_{j=1}^{n} a_{2j} \left[(-1)^{2+j} minor(a_{2j}) \right], \text{along row two}$$

$$|A| = \sum_{j=1}^{n} a_{i2} \left[(-1)^{i+2} minor(a_{i2}) \right], \text{along column two}$$

Again, we can expand along any row or column to obtain the determinant of a matrix. From our definition, it is easily seen that any matrix with a row or a column of zero has $|A| = 0$. These will be illustrated later with examples.

Singular and non-singular matrices: A matrix B is said to be singular if and only if $|B| = 0$, otherwise B is said to be nonsingular.

Invertible Matrix: A matrix B is said to be invertible if and only if it is nonsingular. If B is invertible, then there is a B^{-1} such that

$$BB^{-1} = B^{-1}B = I.$$

Adjoint Matrix: The adjoint matrix of A, denoted as $adjA$, is defined by

$$A(adjA) = (adjA)A = |A|I$$

that is; the adjoint of A is that matrix which multiplies A to give $|A|$I. Operationaly, it is defined as the transpose of the cofactor matrix. The cofactor matrix of A is the matrix in which each a_{ij} is replaced with $cofactor(a_{ij})$. This will be illustrated later with an example.

Matrix algebra

Let M be a collection of matrices over the scalar field K. If A, B and C are non-singular elements of M, and the scalars k_1 and k_2 belong to K, then the following empirical relations hold.

i) $-A$ belongs to M, such that $A + -A = N$, a null matrix.

ii) $A + B = B + A$ *(Addition commutativity)*

iii) $A + (B + C) = (A + B) + C$ *(Addition associativity)*

iv) $A(BC) = (AB)C$ *(Multiplication associativity)*

v) $(k_1A)^{-1} = \frac{1}{k_1}A^{-1}$; $k_1 \neq 0$, $|A| \neq 0$

vi) $(A^{-1})^{-1} = A$

vii) $(AB)^{-1} = B^{-1}A^{-1}$

viii) $(AB)^T = B^TA^T$, $(k_1A)^T = k_1A^T$, $(A^T)^T = A$

ix) $(A + B)^T = A^T + B^T$

x) $A(B - C) = AB - AC$ *(Left distributivity)*

xi) $(B - C)A = BA - CA$ *(Right distributivity)*

xii) $0A = A0 = 0$

xiii) $(k_1 - k_2)A = k_1A - k_2A$

xiv) $k_1(A - B) = k_1A - k_1B$

xv) $A(k_1B) = k_1(AB) = (k_1A)B$

xvi) $(A^{-1})^T = (A^T)^{-1}$

xvii) $(A^k)^T = (A^T)^k$. *Note that* $A^k = A.A.A... , k$ times.

PROPERTIES OF DETERMINANTS

Let M be a collection of matrices over a field K. If A and B are nonsingular nxn elements of M and the scalar k belongs to K, then

i) $|AB| = |BA| = |A||B|$

ii) $|A^T| = |A|$

iii) $|kA| = k^n|A|$

iv) $|A^{-1}| = \dfrac{1}{|A|}$

v) $|A^{-1}A| = |AA^{-1}| = |I_n| = 1$

vi) $|(AB)^{-1}| = |B^{-1}A^{-1}| = |B^{-1}||A^{-1}|$

$$= \frac{1}{|B|}\frac{1}{|A|} = \frac{1}{|A|}\frac{1}{|B|} = |A^{-1}||B^{-1}| = |A^{-1}B^{-1}| = |(BA)^{-1}|$$

vii) $|diag(d_1, d_2, ... d_n)| = d_1 d_2 ... d_n$

viii) If A is an upper or a lower triangular matrix, then the determinant of A is the product of the main - diagonal elements.

PROPERTIES OF ADJOINT

Recall that

$$A(adjA) = (adjA)A = |A|I_n$$
$$\therefore |A(adjA)| = \big||A|I_n\big|$$
$$\Rightarrow (implies\ that)\ |A|\,|adj(A)| = (|A|)^n|I_n|$$
$$|A|\,|adj(A)| = (|A|)^n$$
$$\Rightarrow \quad |adj(A)| = (|A|)^{n-1} \qquad ----P159$$

i)   Again,

$$adjA(adj(adjA)) = |adjA|I_n$$
$$\therefore \big|adjA(adj(adjA))\big| = \big||adjA|I_n\big|$$

10

$$|adjA|\left|\left(adj(adjA)\right)\right| = (|adjA|)^n$$
$$\left|\left(adj(adjA)\right)\right| = (|adjA|)^{n-1}$$
$$\left|\left(adj(adjA)\right)\right| = (|A|^{n-1})^{n-1}$$
$$\left|\left(adj(adjA)\right)\right| = (|A|)^{n^2-2n+1} \qquad ----P160$$

ii)   If $A = adjA$, then $|A| = \pm 1$, $provided\ that |A| \neq 0$

Proof

$$A\ adjA = |A|I_n$$
$$A^2 = |A|I_n$$

Since $A = adjA$, we have

$$|A^2| = \left||A|I_n\right|$$
$$(|A|)^2 = (|A|)^n$$
$$(|A|)^n - (|A|)^2 = 0$$
$$(|A|)^2((|A|)^{n-2} - 1) = 0$$
$$(|A|)^{n-2} = 1$$

Since $|A| \neq 0$, we have

$$\Rightarrow |A| = \pm 1\ ,\ \text{depending on } n$$

If n is odd, then $n - 2$ is odd

$$\therefore (|A|)^{n-2} = 1 \Rightarrow |A| = 1$$

If n is even, then $n - 2$ is even

$$\therefore (|A|)^{n-2} = 1 \Rightarrow |A| = \pm 1$$

This means that if A is an nxn matrix, n is odd and A = adjA , then

$$|A| = 1.$$

ELEMENTARY ROW OPERATIONS: ROW EQUIVALENCE

There are three basic operations called elementary row operations, usually performed on matrices. These are:

i)   *Interchanging any two rows of the matrix*

ii) *Multiplying all the elements of a particular row by the same nonzero scalar*

iii) *Replacing any row in the matrix by its sum with a constant multiple of another row.* By sum of two rows, we mean the addition of the corresponding elements of the two rows.

An *elementary matrix* is defined as a matrix obtained by a single elementary row operation on the identity matrix. For example, $A = \begin{pmatrix} 0 & 1 & 0 \\ 1 & 0 & 0 \\ 0 & 0 & 1 \end{pmatrix}$ is an elementary matrix.

Any two matrices A and B are said to be *row equivalent* if either one can be obtained from the other by a finite sequence of elementary row operations.

Three Elementary Properties of Determinants

i)   If A is a matrix, then the determinant of a matrix B obtained by multiplying a row of A by a scalar k is $k|A|$.

ii) If A is a matrix, then the determinant of a matrix B obtained by interchanging any two rows of A is $-|A|$.

iii) If A is a matrix, then the determinant of a matrix B obtained by replacing a row of A by its sum with a constant multiple of another row is $|A|$; that is, this kind of elementary row operation on A leaves the determinant of A unaltered. In summary,

i.   If $A = \begin{pmatrix} a_{11} & a_{12} & a_{13} \\ a_{21} & a_{22} & a_{23} \\ a_{31} & a_{32} & a_{33} \end{pmatrix}$, and k is any nonzero scalar then,

$$\begin{vmatrix} ka_{11} & ka_{12} & ka_{13} \\ a_{21} & a_{22} & a_{23} \\ a_{31} & a_{32} & a_{33} \end{vmatrix} = \begin{vmatrix} a_{11} & a_{12} & a_{13} \\ ka_{21} & ka_{22} & ka_{23} \\ a_{31} & a_{32} & a_{33} \end{vmatrix} = \begin{vmatrix} a_{11} & a_{12} & a_{13} \\ a_{21} & a_{22} & a_{23} \\ ka_{31} & ka_{32} & ka_{33} \end{vmatrix} = k|A|$$

ii. $$\begin{vmatrix} a_{21} & a_{11} & a_{23} \\ a_{11} & a_{12} & a_{13} \\ a_{31} & a_{32} & a_{33} \end{vmatrix} = -\begin{vmatrix} a_{31} & a_{32} & a_{33} \\ a_{21} & a_{22} & a_{23} \\ a_{11} & a_{12} & a_{13} \end{vmatrix} = -\begin{vmatrix} a_{11} & a_{12} & a_{13} \\ a_{31} & a_{32} & a_{33} \\ a_{21} & a_{22} & a_{23} \end{vmatrix} = -|A|$$

iii.
$$\begin{vmatrix} a_{11} + ka_{31} & a_{12} + ka_{32} & a_{13} + ka_{33} \\ a_{21} & a_{22} & a_{23} \\ a_{31} & a_{32} & a_{33} \end{vmatrix} = |A|$$

It is these three basic properties that account for all the other properties of determinant previously listed; that is, from these three one can easily prove the previous ones.

Combining the third elementary property, and the fact that $|A| = 0$, if A has a row or a column of zero, it is easy to see that any matrix with two identical rows or columns has a determinant equal to zero.

BACK TO SYSTEM OF LINEAR EQUATIONS

Now we are equipped with the tools needed to provide a formal solution to a system of linear equations. Recall that a system of linear equations can be solved by replacing it with an equivalent but simpler one. The concept of row-equivalence matrix helps us to achieve this. Let the equation

$$AX = B$$

represent the matrix form of a system of linear equations, then the matrix
$[A/B]$ or (A/B) formed by concatenating or combining the coefficient matrix A with the constant matrix B is called the associated *augmented matrix* of the system. Simplifying the system means performing the elementary row operations on the augmented matrix until the augmented matrix becomes a *row-echelon matrix* which is always a simpler, yet equivalent representation of the system of linear equations. A matrix is said to be in its row echelon form if:

i)   any zero row is below a nonzero row, a row with at least one nonzero element; a zero row is that whose elements are all zero.

ii)   the first nonzero element of a row is to the right of the first nonzero element of the row immediately above it. For example,

$$A = \begin{pmatrix} 2 & 3 & 0 \\ 0 & 1 & 2 \\ 0 & 0 & 1 \end{pmatrix} \text{ is an echelon matrix}$$

$$B = \begin{pmatrix} 2 & 3 \\ 0 & 1 \\ 0 & 0 \end{pmatrix} \text{ is an echelon matrix}$$

$$C = \begin{pmatrix} 2 & 3 & 1 \\ 0 & 1 & 1 \end{pmatrix} \text{ is an echelon matrix}$$

$$D = \begin{pmatrix} 2 & 0 \\ 0 & 1 \\ 0 & 0 \end{pmatrix} \text{ is an echelon matrix}$$

$$E = \begin{pmatrix} 0 & 2 & 1 & -3 \\ 0 & 0 & 0 & 4 \\ 0 & 0 & 0 & 0 \end{pmatrix} \text{ is an echelon matrix}$$

$$F = \begin{pmatrix} 2 & 3 & 4 \\ 0 & 0 & 0 \\ 1 & 0 & 0 \end{pmatrix} \text{ is not an echelon matrix}$$

$$G = \begin{pmatrix} 2 & 1 & 0 \\ 1 & 3 & 1 \\ 0 & 0 & 0 \end{pmatrix} \text{ is not an echelon matrix}$$

Note that the upper triangular matrix is always an echelon matrix. Let

$$[A/B] = \begin{pmatrix} a_{11} & a_{12} & a_{13} & b_1 \\ a_{21} & a_{22} & a_{23} & b_2 \\ a_{31} & a_{32} & a_{33} & b_3 \end{pmatrix} \text{ be the augmented matrix of the linear system}$$

$$a_{11}x_1 + a_{12}x_2 + a_{13}x_3 = b_1$$
$$a_{21}x_1 + a_{22}x_2 + a_{23}x_3 = b_2$$
$$a_{31}x_1 + a_{32}x_2 + a_{33}x_3 = b_3$$

If the row echelon form

$$\begin{pmatrix} u_{11} & u_{12} & u_{13} & v_1 \\ 0 & u_{22} & u_{23} & v_2 \\ 0 & 0 & u_{33} & v_3 \end{pmatrix}$$ is obtained from $[A/B]$ upon a finite sequence of elementary

row operations, then the linear system

$$u_{11}x_1 + u_{12}x_2 + u_{13}x_3 = v_1$$
$$u_{22}x_2 + u_{23}x_3 = v_2$$
$$u_{33}x_3 = v_3$$

is equivalent in all respects to the previous system. This means that the two systems have the same solution. However, it is easier to solve the system in the echelon form than in its given form. Solving the echelon system, therefore, we have

$$x_3 = \frac{v_3}{u_{33}}, \ x_2 = v_2 - \frac{\frac{u_{23}v_3}{u_{33}}}{u_{22}}, \ x_1 = \frac{v_1 - u_{13}x_3 - u_{12}x_2}{u_{11}}.$$ This is the **Gaussian**

Elimination method.

Example 1.1:

Using the Gaussian elimination method, solve the system

$$2x_1 - 3x_2 = 8$$
$$3x_1 + 5x_2 = 31$$

Solution

The augmented matrix of the system is

$$A = \begin{pmatrix} 2 & -3 & 8 \\ 3 & 5 & 31 \end{pmatrix}$$

Multiplying row 1 by 1/2 , we have

$$\begin{pmatrix} 2 & -3 & 8 \\ 3 & 5 & 31 \end{pmatrix} \sim \begin{pmatrix} 1 & \frac{-3}{2} & 4 \\ 3 & 5 & 31 \end{pmatrix} = A(\ \tfrac{1}{2}\,R_1) = A_1$$

Replacing row 2 (subsequently) by its sum with -3 times row 1, we have

$$A_2 = A_1(R_2 - 3R_1) = \begin{pmatrix} 1 & \dfrac{-3}{2} & 4 \\ 0 & 5 + \dfrac{9}{2} & 31 - 12 \end{pmatrix}$$

The matrices A, A_1 and A_2 are all equivalent since they are related by a finite sequence of elementary row operations. It is conventional to denote equivalence by the symbol " $\sim$ ". Therefore,

$$A \sim A_1 \sim A_2$$

$$\begin{pmatrix} 2 & -3 & 8 \\ 3 & 5 & 21 \end{pmatrix} \sim \begin{pmatrix} 1 & \dfrac{-3}{2} & 4 \\ 0 & \dfrac{19}{2} & 19 \end{pmatrix}$$

Since the last augmented matrix A_2 describes the same system as the augmented matrix A, due to equivalence; therefore,

$$\frac{19}{2} x_2 = 19$$

$$x_2 = 2 \text{ , and}$$

$$x_1 - \frac{3x_2}{2} = 4$$

$$x_1 = 7$$

Therefore, the ordered pair $(7, 2)$ or the column vector $\begin{pmatrix} 7 \\ 2 \end{pmatrix}$ is the solution to the system.

Example1.2:

Solve the system

$$3x_1 - x_2 + 2x_3 = -1$$
$$-x_1 + 5x_2 + 10x_3 = 13$$
$$2x_1 - 3x_2 - 8x_3 = -3$$

Solution

$$\begin{pmatrix} 3 & -1 & 2 & -1 \\ -1 & 5 & 10 & 13 \\ 2 & -3 & -8 & -3 \end{pmatrix} (R_3 + 2R_2) \sim \begin{pmatrix} 3 & -1 & 2 & -1 \\ -1 & 5 & 10 & 13 \\ 0 & 7 & 12 & 23 \end{pmatrix}$$

$$\tfrac{1}{3}R_1 \sim \begin{pmatrix} 1 & \tfrac{-1}{3} & \tfrac{2}{3} & \tfrac{-1}{3} \\ -1 & 5 & 10 & 13 \\ 0 & 7 & 12 & 23 \end{pmatrix} (R_2 + R_1) \sim \begin{pmatrix} 1 & \tfrac{-1}{3} & \tfrac{2}{3} & \tfrac{-1}{3} \\ 0 & \tfrac{14}{3} & \tfrac{32}{3} & \tfrac{38}{3} \\ 0 & 7 & 12 & 23 \end{pmatrix} \tfrac{2}{3}R_3$$

$$\begin{pmatrix} 1 & \tfrac{-1}{3} & \tfrac{2}{3} & \tfrac{-1}{3} \\ 0 & \tfrac{14}{3} & \tfrac{32}{3} & \tfrac{38}{3} \\ 0 & \tfrac{14}{3} & 8 & \tfrac{46}{3} \end{pmatrix} (R_3 - R_2) \sim \begin{pmatrix} 1 & \tfrac{-1}{3} & \tfrac{2}{3} & \tfrac{-1}{3} \\ 0 & \tfrac{14}{3} & \tfrac{32}{3} & \tfrac{38}{3} \\ 0 & 0 & \tfrac{-8}{3} & \tfrac{8}{3} \end{pmatrix}$$

$$\therefore \begin{pmatrix} 3 & -1 & 2 & -1 \\ -1 & 5 & 10 & 13 \\ 2 & -3 & -8 & -3 \end{pmatrix} \sim \begin{pmatrix} 1 & \tfrac{-1}{3} & \tfrac{2}{3} & \tfrac{-1}{3} \\ 0 & \tfrac{14}{3} & \tfrac{32}{3} & \tfrac{38}{3} \\ 0 & 0 & \tfrac{-8}{3} & \tfrac{8}{3} \end{pmatrix}$$

$$\tfrac{-8}{3}x_3 = \tfrac{8}{3}, \; x_3 = -1$$

$$\tfrac{14}{3}x_2 + \tfrac{32}{3}x_3 = \tfrac{38}{3}$$

$$14x_2 + 32(-1) = 38$$

$$x_2 = \frac{70}{14} = 5$$

$$x_1 - \frac{1}{3}x_2 + \frac{2}{3}x_3 = \frac{-1}{3}$$

$$3x_1 = -1 + x_2 - 2x_3$$

$$x_1 = \frac{6}{3} = 2$$

The solution vector is equal to $\begin{pmatrix} x_1 \\ x_2 \\ x_3 \end{pmatrix} = \begin{pmatrix} 2 \\ 5 \\ -1 \end{pmatrix}$

Note: we could have avoided the unnecessary fractions by interchanging rows 2 and 1 somewhere along the sequence of row operations; also, note that $R_i + kR_j$ conventionally means replace row i with its sum with k multiplied by row j.

Example 1.3:

Solve the system

$3x_1 - x_2 + x_3 = 4$

$x_1 + x_2 - x_3 = 4$

$x_1 - 2x_2 + 3x_3 = 4$

Solution

$$\begin{bmatrix} 3 & -1 & 1 & | & 4 \\ 1 & 1 & -1 & | & 4 \\ 1 & -2 & 3 & | & -5 \end{bmatrix} (R_2 \sim R_1) \sim \begin{bmatrix} 1 & 1 & -1 & | & 4 \\ 3 & -1 & 1 & | & 4 \\ 1 & -2 & 3 & | & -5 \end{bmatrix} R_3 - R_1 \sim$$

$$\begin{bmatrix} 1 & 1 & -1 & | & 4 \\ 3 & -1 & 1 & | & 4 \\ 0 & -3 & 4 & | & -9 \end{bmatrix} (R_2 - 3R_1) \sim \begin{bmatrix} 1 & 1 & -1 & | & 4 \\ 0 & -4 & 4 & | & -8 \\ 0 & -3 & 4 & | & -9 \end{bmatrix} \frac{1}{4} R_2 \sim$$

$$\begin{bmatrix} 1 & 1 & -1 & | & 4 \\ 0 & -1 & 1 & | & -2 \\ 0 & -3 & 4 & | & -9 \end{bmatrix} (R_3 - 3R_2) \sim \begin{bmatrix} 1 & 1 & -1 & | & 4 \\ 0 & -1 & 1 & | & -2 \\ 0 & 0 & 1 & | & -3 \end{bmatrix}$$

$$\therefore \begin{bmatrix} 3 & -1 & 1 & | & 4 \\ 1 & 1 & -1 & | & 4 \\ 1 & -2 & 3 & | & -5 \end{bmatrix} \sim \begin{bmatrix} 1 & 1 & -1 & | & 4 \\ 0 & -1 & 1 & | & -2 \\ 0 & 0 & 1 & | & -3 \end{bmatrix}$$

[**Note:** $(R_i \sim R_j)$ means interchange rows i and j.]

$$x_3 = -3$$

$$-x_2 + x_3 = -2, \ x_2 = -1$$

$$x_1 + x_2 - x_3 = 4, \ x_1 = 2$$

Therefore, the solution vector is

$$\begin{pmatrix} x_1 \\ x_2 \\ x_3 \end{pmatrix} = \begin{pmatrix} 2 \\ -1 \\ -3 \end{pmatrix}$$

We could have gone further to reduce the last augmented matrix

$$\begin{bmatrix} 1 & 1 & -1 & | & 4 \\ 0 & -1 & 1 & | & -2 \\ 0 & 0 & 1 & | & -3 \end{bmatrix} \overset{(R_1 + R_2)}{\sim} \begin{bmatrix} 1 & 1 & 0 & | & 2 \\ 0 & -1 & 1 & | & -2 \\ 0 & 0 & 1 & | & -3 \end{bmatrix} (R_2 - R_3)$$

$$\begin{bmatrix} 1 & 0 & 0 & | & 2 \\ 0 & -1 & 0 & | & 1 \\ 0 & 0 & 1 & | & -3 \end{bmatrix} \overset{(-R_2)}{\sim} \begin{bmatrix} 1 & 0 & 0 & | & 2 \\ 0 & 1 & 0 & | & -1 \\ 0 & 0 & 1 & | & -3 \end{bmatrix}$$

Again,

$$\begin{bmatrix} 3 & -1 & 1 & | & 4 \\ 1 & 1 & -1 & | & 4 \\ 1 & -2 & 3 & | & -5 \end{bmatrix} \sim \begin{bmatrix} 1 & 0 & 0 & | & 2 \\ 0 & 1 & 0 & | & -1 \\ 0 & 0 & 1 & | & -3 \end{bmatrix}$$

$$\Rightarrow \quad x_3 = -3 \,, x_2 = -1 \,, \ x_1 = 2$$

$$\begin{pmatrix} x_1 \\ x_2 \\ x_3 \end{pmatrix} = \begin{pmatrix} 2 \\ -1 \\ -3 \end{pmatrix}, \text{ the solution vector.}$$

This further reduced echelon form is called *reduced-row echelon form.* The process of reducing a matrix, using elementary row operations into its reduced-row echelon form is called **Gauss-Jordan reduction.**

Example 1.4:

Solve the system

$x_1 - x_2 + 2x_3 - x_4 = 6$

$2x_1 + 3x_2 + 4x_3 - 2x_4 = 7$

$-x_1 + x_2 + 2x_3 + x_4 = -4$

$6x_1 + 2x_2 + x_4 = 4$

Solution

$$\begin{bmatrix} 1 & -1 & 2 & -1 & 6 \\ 2 & 3 & 4 & -2 & 7 \\ -1 & 1 & 2 & 1 & -4 \\ 6 & 2 & 0 & 1 & 1 \end{bmatrix} \overset{(R_2 - R_3)}{\sim} \begin{bmatrix} 1 & -1 & 2 & -1 & 6 \\ -1 & 1 & 2 & 1 & -4 \\ 2 & 3 & 4 & -2 & 7 \\ 6 & 2 & 0 & 1 & 1 \end{bmatrix}$$

$$\overset{(R_4 - 3R_3)}{\sim} \begin{bmatrix} 1 & -1 & 2 & -1 & 6 \\ -1 & 1 & 2 & 1 & -4 \\ 2 & 3 & 4 & -2 & 7 \\ 0 & -7 & -12 & 7 & -20 \end{bmatrix} \overset{(R_3 + 2R_2)}{\sim} \begin{bmatrix} 1 & -1 & 2 & -1 & 6 \\ -1 & 1 & 2 & 1 & -4 \\ 0 & 5 & 8 & 0 & -1 \\ 0 & -7 & -12 & 7 & -20 \end{bmatrix}$$

$$\overset{(R_2 + R_1)}{\sim} \begin{bmatrix} 1 & -1 & 2 & -1 & 6 \\ 0 & 0 & 4 & 0 & 2 \\ 0 & 5 & 8 & 0 & -1 \\ 0 & -7 & -12 & 7 & -20 \end{bmatrix} \overset{(R_4 + 3R_2)}{\sim} \begin{bmatrix} 1 & -1 & 2 & -1 & 6 \\ 0 & 0 & 4 & 0 & 2 \\ 0 & 5 & 8 & 0 & -1 \\ 0 & -7 & -12 & 7 & -14 \end{bmatrix}$$

$$\overset{(R_3 - 2R_2)(\frac{1}{4}R_4)}{\sim} \begin{bmatrix} 1 & -1 & 2 & -1 & 6 \\ 0 & 0 & 4 & 0 & 2 \\ 0 & 5 & 0 & 0 & -5 \\ 0 & -1 & 0 & 1 & -2 \end{bmatrix} \overset{(\frac{1}{4}R_2)\,(\frac{1}{5}R_3)}{\sim}$$

$$\begin{bmatrix} 1 & -1 & 2 & -1 & 6 \\ 0 & 0 & 1 & 0 & \frac{1}{2} \\ 0 & 1 & 0 & 0 & -1 \\ 0 & -1 & 0 & 1 & -2 \end{bmatrix} \overset{(R_1 - 2R_2)}{\sim} \begin{bmatrix} 1 & -1 & 0 & -1 & 5 \\ 0 & 0 & 1 & 0 & \frac{1}{2} \\ 0 & 1 & 0 & 0 & -1 \\ 0 & -1 & 0 & 1 & -2 \end{bmatrix}$$

$$(R_1 + R_3),(R_4 + R_3) \overset{}{\sim} \begin{bmatrix} 1 & 0 & 0 & -1 & 4 \\ 0 & 0 & 1 & 0 & \frac{1}{2} \\ 0 & 1 & 0 & 0 & -1 \\ 0 & 0 & 0 & 1 & -3 \end{bmatrix}$$

$$(R_1 + R_4),(R_2 \sim R_3) \overset{}{\sim} \begin{bmatrix} 1 & 0 & 0 & 0 & 1 \\ 0 & 1 & 0 & 0 & -1 \\ 0 & 0 & 1 & 0 & \frac{1}{2} \\ 0 & 0 & 0 & 1 & -3 \end{bmatrix}$$

$$\Rightarrow \begin{bmatrix} x_1 \\ x_2 \\ x_3 \\ x_4 \end{bmatrix} = \begin{bmatrix} 1 \\ -1 \\ \frac{1}{2} \\ -3 \end{bmatrix}$$

Note: We have shortened the process by combining two simple steps in one step; one can even take more than two steps at a time, provided they are simple enough-clear enough to be combined.

Example 1.5:

Solve the system

$3x_1 - x_2 + x_4 = 4$

$-x_1 + 3x_2 - x_3 = -2$

$x_1 + x_2 + 2x_3 + x_4 = 8$

$5x_1 + 3x_2 - 2x_3 + 2x_4 = 5$

Solution

$$\begin{pmatrix} 3 & -1 & 0 & 1 & | & 4 \\ -1 & 3 & -1 & 0 & | & -2 \\ 1 & 1 & 2 & 1 & | & 8 \\ 5 & 3 & -2 & 2 & | & 5 \end{pmatrix} (R_3 \sim R_1) \begin{pmatrix} 1 & 1 & 2 & 1 & | & 8 \\ -1 & 3 & -1 & 0 & | & -2 \\ 3 & -1 & 0 & 1 & | & 4 \\ 5 & 3 & -2 & 2 & | & 5 \end{pmatrix}$$

$$(R_4 - (R_3 + 2R_1)) \begin{pmatrix} 1 & 1 & 2 & 1 & | & 8 \\ -1 & 3 & -1 & 0 & | & -2 \\ 3 & -1 & 0 & 1 & | & 4 \\ 0 & 2 & -6 & -1 & | & -15 \end{pmatrix} \begin{matrix} (R_3 - 3R_1) \\ (R_1 + R_2) \end{matrix} \begin{pmatrix} 0 & 4 & 1 & 1 & | & 6 \\ -1 & 3 & -1 & 0 & | & -2 \\ 0 & -4 & -6 & -2 & | & -20 \\ 0 & 2 & -6 & -1 & | & -15 \end{pmatrix}$$

$$(R_4 - R_3), (R_3 + R_1) \sim \begin{pmatrix} 0 & 4 & 1 & 1 & | & 6 \\ -1 & 3 & -1 & 0 & | & -2 \\ 0 & 0 & -5 & -1 & | & -14 \\ 0 & 6 & 0 & 1 & | & 5 \end{pmatrix} \begin{matrix} (R_1 + R_3) \\ (\frac{1}{4}R_1) \end{matrix} \sim$$

$$\begin{pmatrix} 0 & 1 & -1 & 0 & | & -2 \\ -1 & 3 & -1 & 0 & | & -2 \\ 0 & 0 & -5 & -1 & | & -14 \\ 0 & 6 & 0 & 1 & | & 5 \end{pmatrix}$$

$(R_3-5R_1),(R_4+R_3)$ $\begin{pmatrix} 0 & 1 & -1 & 0 & | & -2 \\ -1 & 3 & -1 & 0 & | & -2 \\ 0 & -5 & 0 & -1 & | & -4 \\ 0 & 1 & 0 & 0 & | & 1 \end{pmatrix}$ $\begin{matrix} (R_1-R_4) \\ (R_2-R_1) \end{matrix}$ $\begin{pmatrix} 0 & 0 & -1 & 0 & | & -3 \\ -1 & 3 & 0 & 0 & | & 1 \\ 0 & -5 & 0 & -1 & | & -4 \\ 0 & 1 & 0 & 0 & | & 1 \end{pmatrix}$

$(R_2-3R_4),(R_3+5R_4)$ $\begin{pmatrix} 0 & 0 & -1 & 0 & | & -3 \\ -1 & 0 & 0 & 0 & | & -2 \\ 0 & 0 & 0 & -1 & | & 1 \\ 0 & 1 & 0 & 0 & | & 1 \end{pmatrix}$

$\begin{matrix} (-R_1) \\ (-R_2) \\ (-R_3) \\ (R_1\sim R_2) \\ (R_4\sim R_2) \\ (R_3\sim R_4) \end{matrix}$ $\sim$ $\begin{pmatrix} 1 & 0 & 0 & 0 & | & 2 \\ 0 & 1 & 0 & 0 & | & 1 \\ 0 & 0 & 1 & 0 & | & 3 \\ 0 & 0 & 0 & 1 & | & -1 \end{pmatrix}$

$$\Rightarrow \begin{pmatrix} x_1 \\ x_2 \\ x_3 \\ x_4 \end{pmatrix} = \begin{pmatrix} 2 \\ 1 \\ 3 \\ -1 \end{pmatrix}$$

TEST FOR THE EXISTENCE OF A SOLUTION TO A LINEAR SYSTEM: MATRIX RANK

It is not always the case that a linear system has a solution; even if it has a solution, the solution may not be *unique*; that is, it may have more than one solution. For instance, the linear system

$$x + y = 8$$
$$2x + 2y = 16$$

has more than one solution. This means that the solution is not unique. This particular system has no unique solution because we do not have enough information about it as the second equation is just a scaled version of the first – the two equations are the same. This is because each of them always implies the other. Again, the linear system

$$x + y = 8$$
$$2x + 2y = 15$$

has no solution because there is no vector $\begin{pmatrix} x \\ y \end{pmatrix}$ that satisfies it. This is because the system is *inconsistent*. The left hand side of the second equation is actually double that of the first-meaning that the right hand side of the second equation is a fixed number, given the first one; we are not free to choose it. Since the RHS of the second equation is different from its fixed value, then the two equations are not in agreement, and thus there is inconsistency; no solution vector exists for it.

The above two cases of *nonuniqueness* and *inconsistency* are simple enough to be detected without performing any test or calculation. As more equations enter into the system, weather as a result of an increase in the number of unknown or due to redundancy (more equations than number of unknowns), things become more involved; so, it becomes more difficult to tell at single glance the system characteristics. Now let us consider a formal and systematic way of characterizing a system of linear equations.

THE RANK OF MATRIX

The rank of a matrix A is defined as the order of the largest nonsingular submatrix that can be found in matrix A. Every matrix is a submatrix of itself, and so when considering the largest submatrix of a matrix, the matrix itself is considered first, unless it is not a square matrix. Again, we need only consider square sub matrices because the terms "singular" and "nonsingular" apply to square matrices only.

Example 1.6:

Determine the rank of the following matrices

a) $\begin{pmatrix} 2 & 4 \\ 1 & 2 \end{pmatrix}$ (b) $\begin{pmatrix} 2 & -1 \\ 1 & 1 \end{pmatrix}$ (c) $\begin{pmatrix} 3 & 2 & 4 \\ -1 & 1 & 3 \\ -1 & 6 & 16 \end{pmatrix}$ (d) $\begin{pmatrix} 2 & 3 & 1 \\ -2 & 4 & -1 \\ 1 & 0 & 5 \end{pmatrix}$

Solution

We simply want to determine the order of the largest nonsingular sub matrix of each of the matrices. A nonsingular matrix is that whose determinant is not equal to zero. This means our job is equivalent to evaluating determinants.

For (a)

$$A = \begin{pmatrix} 2 & 4 \\ 1 & 2 \end{pmatrix} ; \text{ we start from the matrix itself.}$$

$$|A| = \begin{vmatrix} 2 & 4 \\ 1 & 2 \end{vmatrix} = 2(2) - 4(1) = 0$$

$$\Rightarrow rank < 2 ; \therefore rank \text{ is 1.}$$

This is because the next lower submatrices of A are the singleton matrices (2), (4), (1), (2), which are obviously nonsingular 1x1 matrices because the determinant of a singleton matrix $A = (a)$ is $|A| = a$, the number itself. Only the null matrix has a rank of zero. This means that any 2x2 singular matrix other than a null matrix is of rank 1. Therefore, $rank (A) = 1$.

Note that we need only one of the four entries of A to be different from zero.

b) $\qquad A = \begin{pmatrix} 2 & -1 \\ 1 & 1 \end{pmatrix} ; |A| = 2(1) - 1(-1) = 4$

$$rank (A) = 2$$

Please don't misinterpret the definition, the rank of an nxn nonsingular matrix is n, not n^2.

c) $\qquad A = \begin{pmatrix} 3 & 2 & 4 \\ -1 & 1 & 3 \\ -1 & 6 & 6 \end{pmatrix}$

Recall the general formula and the definition of a determinant.

$$det A = |A| = \sum_{j=1}^{3} a_{1j} \, cofactor \, (a_{1j}) \, ; \text{ expanding along the first row}$$

$$\text{Where } cofactor \, (a_{1j}) = (-1)^{i+j} minor \, (a_{ij}).$$

$$cofactor(a_{11}) = (-1)^{1+1} \begin{vmatrix} 1 & 3 \\ 6 & 16 \end{vmatrix} = -2 = cofactor \, (3)$$

$$cofactor(a_{12}) = (-1)^{1+2} \begin{vmatrix} -1 & 3 \\ -6 & 16 \end{vmatrix} = 13 = cofactor(2)$$

$$cofactor(a_{13}) = (-1)^{1+3} \begin{vmatrix} -1 & 3 \\ -6 & 16 \end{vmatrix} = 5 = cofactor(4)$$

$$|A| = \sum_{j=1}^{3} a_{1j} \, cofactor(a_{1j}) = 3(-2) + 2(13) + 4(-5)$$

$$|A| = 0 \Rightarrow rank \, (A) < 3$$

Therefore we go down further to 2 x 2 sub matrices of A. We need at least one of them to be nonsingular in order to conclude that $rank(A) = 2$. $Rank(A)$ can only be equal to 1 if and only if all the possible 2x2 submatrices are singular, seeing that A is not a null matrix. It is obvious that the submatrix $\begin{pmatrix} 2 & 4 \\ 1 & 3 \end{pmatrix}$ is nonsingular, that is

$$\begin{vmatrix} 2 & 4 \\ 1 & 3 \end{vmatrix} \neq 0. \quad \therefore rank(A) = 2$$

d.) $\quad A = \begin{pmatrix} 2 & 3 & 1 \\ -2 & 4 & -1 \\ 1 & 0 & 5 \end{pmatrix}$, expanding along the second column, we have

$$|A| = 3(-1)^{1+2} \, (-10 + 1) + 4(-1)^{2+2} \, (10 - 1) + 0.(-1)^{3+2} \, (-2 + 2)$$

$$|A| = 27 + 36 + 0 \neq 0$$

$$\Rightarrow rank(A) = 3$$

It is instructive to let you know that the above method that evaluates determinant to determine the rank of a matrix is not the formal way. The formal method is: reduction of the matrix to its row-echelon form. The rank of the matrix, after reducing to its row-echelon form, is the number of nonzero rows found in it.

Example 1.7:

Determine the rank of the following matrices:

(a.) $\begin{pmatrix} 2 & 3 & 1 & 4 \\ 1 & 2 & 3 & 1 \\ 3 & 5 & 4 & 5 \end{pmatrix}$ (b.) $\begin{pmatrix} 2 & 1 & 1 \\ 1 & 1 & -1 \\ 4 & 2 & 0 \end{pmatrix}$ (c.) $\begin{pmatrix} 1 & 2 & 1 \\ 3 & 0 & 2 \end{pmatrix}$

Solution

We can still use the method of determinant, even though (a) and (c) are not square matrices. All we need do is consider the largest square submatrix and then find its determinant. That is

a) $A = \begin{pmatrix} 2 & 3 & 1 & 4 \\ 1 & 2 & 3 & 2 \\ 3 & 5 & 4 & 5 \end{pmatrix}$, the matrix $\begin{pmatrix} 2 & 3 & 1 \\ 1 & 2 & 3 \\ 3 & 5 & 4 \end{pmatrix}$ is a largest square submatrix of A.

$$\begin{vmatrix} 2 & 3 & 1 \\ 1 & 2 & 3 \\ 3 & 5 & 4 \end{vmatrix} = 0$$

Test for yourself that every other 3 x 3 submatrix of A is singular, but

$$\begin{vmatrix} 2 & 3 \\ 1 & 2 \end{vmatrix} \neq 0$$

$$rank(A) = 2$$

You can imagine how tedious it is to evaluate several 3x3 determinants. This is where the row-echelon method comes in. Now we use the row-echelon method on the same matrix.

$$\begin{pmatrix} 2 & 3 & 1 & 4 \\ 1 & 2 & 3 & 1 \\ 3 & 5 & 4 & 5 \end{pmatrix} (R_1 \sim R_2) \begin{pmatrix} 1 & 2 & 3 & 1 \\ 2 & 3 & 1 & 4 \\ 3 & 5 & 4 & 5 \end{pmatrix}$$

$$(R_3 - (R_2 + R_1)) \begin{pmatrix} 1 & 2 & 3 & 1 \\ 2 & 3 & 1 & 4 \\ 0 & 0 & 0 & 0 \end{pmatrix} (R_2 - 2R_1) \begin{pmatrix} 1 & 2 & 3 & 1 \\ 0 & -1 & -5 & 2 \\ 0 & 0 & 0 & 0 \end{pmatrix}$$

The row- echelon form $\begin{pmatrix} 1 & 2 & 3 & 1 \\ 0 & -1 & -5 & 2 \\ 0 & 0 & 0 & 0 \end{pmatrix}$ has two nonzero rows.

$$\therefore rank(A) = 2.$$

b) $\qquad A = \begin{pmatrix} 2 & 1 & 1 \\ 1 & 1 & -1 \\ 4 & 2 & 0 \end{pmatrix} (R_1 \sim R_2) \begin{pmatrix} 1 & 1 & -1 \\ 2 & 1 & 1 \\ 4 & 2 & 0 \end{pmatrix} (R_2 - 2R_1), (R_3 - 4R_1)$

$$\begin{pmatrix} 1 & 1 & -1 \\ 0 & -1 & 3 \\ 0 & -2 & 4 \end{pmatrix} (R_3 - 2R_2) \begin{pmatrix} 1 & 1 & -1 \\ 0 & -1 & 3 \\ 0 & 0 & -2 \end{pmatrix}$$

Since the echelon form $\begin{pmatrix} 1 & 1 & -1 \\ 0 & -1 & 3 \\ 0 & 0 & -2 \end{pmatrix}$ has three nonzero rows.

$$\therefore rank(A) = 3.$$

c) $\quad A = \begin{pmatrix} 1 & 2 & 1 \\ 3 & 0 & 2 \end{pmatrix} (R_2 - 3R_1) \begin{pmatrix} 1 & 2 & 1 \\ 0 & -6 & -1 \end{pmatrix}$, which is in echelon form.

$$\therefore rank(A) = 2.$$

Now let us see how we can use the rank of a matrix to determine whether a system of linear equations has no solution (inconsistent), has only one solution (unique), or has more than one solution (not unique). Consider, therefore, the linear system

$$AX = B$$

All we want to do is to compare the rank of the coefficient matrix $A(nxn)$ and that of the augmented matrix $[A/B]$.

i)   *If $rank(A) < rank([A/B])$, the system is **inconsistent**, and thus has no solution.*

ii) *If $rank(A) = rank([A/B]) = n$, the system has exactly one solution; that is, the system is **unique**.*

iii) *If $rank(A) = rank([A/B]) < n$, the system has an infinite number of solutions; that is, the system is **consistent but not unique**.*

Be informed that any linear system either has no solution, or has exactly one solution, or has an infinite number of solutions. In other words, if a linear system is known to have two solutions or more, then, of necessity, it has infinitely many solutions. To show this, let X_1 and X_2 be two distinct solutions to the system

$$AX = B \qquad\qquad --*$$

Therefore, $AX_1 = B$, and $AX_2 = B$

$$AX_2 - AX_1 = B - B$$
$$A(X_2 - X_1) = 0$$

Let

$$X_2 - X_1 = X_s$$
$$AX_s = 0$$

This implies that X_s is a solution to the *homogeneous system $AX = 0$*

We claim that the vector $X_2 + tX_s$ is also a solution; that is,

$$A(X_2 + tX_s) = 0 \ , \ t \text{ is real.}$$

Because

$$A(X_2 + tX_s) = AX_2 + tAX_s$$
$$= B + t(0) = B$$

This implies that the vector $X_2 + tX_s = X_2 + t(X_2 - X_1)$ is always a solution to equation * whenever X_1 and X_2 are. Since t is any real number, therefore equation * has an infinite number of solutions.

Example 1.8:

Characterize the following systems

a) $x_1 + 3x_2 - 2x_3 = 6$
    $4x_1 + 5x_2 + 2x_3 = 3$
    $x_1 + 3x_2 + 4x_3 = 7$

b) $x_1 + 2x_2 - 4x_3 = 3$
    $x_1 + 2x_2 + 3x_3 = -4$
    $2x_1 + 4x_2 + x_3 = -3$

c) $2x_1 + 3x_2 - x_3 = 1$
    $3x_1 - x_2 + x_3 = 2$
    $x_1 - 4x_2 + 2x_3 = 3$

Solution

a)

$$A = \begin{pmatrix} 1 & 3 & -2 \\ 4 & 5 & 2 \\ 1 & 3 & 4 \end{pmatrix}$$

$$[A/B] = \begin{pmatrix} 1 & 3 & -2 & 6 \\ 4 & 5 & 2 & 3 \\ 1 & 3 & 4 & 7 \end{pmatrix}$$

Let us first determine the rank of A. It is wise to do so because A is part of $[A/B]$. If $rank(A) = 3$, then we do not need to determine $rank([A/B])$; $rank([A/B])$ must be equal to 3 because A is part of $[A/B]$.

$$\begin{pmatrix} 1 & 3 & -2 \\ 4 & 5 & 2 \\ 1 & 3 & 4 \end{pmatrix} (R_2 \sim R_3) \begin{pmatrix} 1 & 3 & -2 \\ 1 & 3 & 4 \\ 4 & 5 & 2 \end{pmatrix} (R_2 - R_1), (R_3 - 4R_1) \begin{pmatrix} 1 & 3 & -2 \\ 0 & 0 & 6 \\ 0 & -7 & 10 \end{pmatrix}$$

$$(R_2 \sim R_3) \begin{pmatrix} 1 & 3 & -2 \\ 0 & -7 & 10 \\ 0 & 0 & 6 \end{pmatrix}$$

$$rank(A) = 3$$

$$\Rightarrow rank([A/B]) = 3$$

Since $rank(A) = rank([A/B]) = 3$, therefore the system has only one solution.

b) $\qquad A = \begin{pmatrix} 1 & 2 & -4 \\ 1 & 2 & 3 \\ 2 & 4 & 1 \end{pmatrix}, [A/B] = \begin{pmatrix} 1 & 2 & -4 & 3 \\ 1 & 2 & 3 & -4 \\ 2 & 4 & 1 & -3 \end{pmatrix}$

$$\begin{pmatrix} 1 & 2 & -4 \\ 1 & 2 & 3 \\ 2 & 4 & 1 \end{pmatrix} (R_3 - (R_2 + R_1)) \begin{pmatrix} 1 & 2 & -4 \\ 1 & 2 & 3 \\ 0 & 0 & 2 \end{pmatrix} (R_2 - R_1) \begin{pmatrix} 1 & 2 & -4 \\ 0 & 0 & 7 \\ 0 & 0 & 2 \end{pmatrix}$$

$$(R_3 - \tfrac{2}{7}R_2) \begin{pmatrix} 1 & 2 & -4 \\ 0 & 0 & 7 \\ 0 & 0 & 0 \end{pmatrix}$$

$$rank(A) = 2$$

We need to determine $rank([A/B])$.

$$[A/B] = \begin{pmatrix} 1 & 2 & -4 & 3 \\ 1 & 2 & 3 & -4 \\ 2 & 4 & 1 & -3 \end{pmatrix} \underset{(R_3-(R_2+R_1))}{\sim} \begin{pmatrix} 1 & 2 & -4 & 3 \\ 1 & 2 & 3 & -4 \\ 0 & 0 & 2 & -2 \end{pmatrix}$$

$$\underset{(R_2-R_1)}{\sim} \begin{pmatrix} 1 & 2 & -4 & 3 \\ 0 & 0 & 7 & -7 \\ 0 & 0 & 2 & -2 \end{pmatrix} \underset{(R_3-\frac{2}{7}R_2)}{\sim} \begin{pmatrix} 1 & 2 & -4 & 3 \\ 0 & 0 & 7 & -7 \\ 0 & 0 & 0 & 0 \end{pmatrix}$$

$$rank([A/B]) = 2$$

Since $rank(A) = rank([A/B]) < 3$, the system has an infinite number of solutions.

b) $\qquad A = \begin{pmatrix} 2 & 3 & -1 \\ 3 & -1 & 1 \\ 1 & -4 & 2 \end{pmatrix}$, $\quad (A/B) = \begin{pmatrix} 2 & 3 & -1 & 1 \\ 3 & -1 & 1 & 2 \\ 1 & -4 & 2 & 3 \end{pmatrix}$

$$\begin{pmatrix} 2 & 3 & 1 \\ 3 & -1 & 1 \\ 1 & -4 & 2 \end{pmatrix} \underset{(R_2\sim R_3),\ (R_1\sim R_2)}{\sim} \begin{pmatrix} 1 & -4 & 2 \\ 2 & 3 & -1 \\ 3 & -1 & 1 \end{pmatrix}$$

$$\underset{(R_3-(R_2+R_1)),\ (R_2-2R_1)}{\sim} \begin{pmatrix} 1 & -4 & 2 \\ 0 & 11 & -5 \\ 0 & 0 & 0 \end{pmatrix}.$$

$$rank(A) = 2$$

We need to determine $rank([A/B])$.

$$\begin{pmatrix} 2 & 3 & -1 & 1 \\ 3 & -1 & 1 & 2 \\ 1 & -4 & 2 & 3 \end{pmatrix} \underset{(R_2\sim R_3),\ (R_1\sim R_2)}{\sim} \begin{pmatrix} 1 & -4 & 2 & 3 \\ 2 & 3 & -1 & 1 \\ 3 & -1 & 1 & 2 \end{pmatrix} \underset{(R_3-(R_2+R_1)),\ (R_2+2R_1)}{\sim}$$

$$\begin{pmatrix} 1 & -4 & 2 & 3 \\ 0 & 11 & -5 & -5 \\ 0 & 0 & 0 & -2 \end{pmatrix}.$$

$$rank([A/B]) = 3$$

Since $rank(A) < rank([A/B])$, the system is inconsistence, and thus has no solution.

Example 1.9:

Characterize the following systems

a) $2x_1 + 5x_2 = 3$
    $3x_1 - x_2 = 1$
    $4x_1 + x_2 = -1$

b) $x_1 - x_2 = 3$
    $2x_1 + 3x_2 = 5$
    $x_1 - 6x_2 = 4$

c) $3x_1 + x_2 - x_3 = 0$
    $-x_1 + 2x_2 + x_3 = 0$
    $2x_1 - x_2 + 2x_3 = 0$

Solution:

The matrix form of the system is

$$\begin{pmatrix} 2 & 5 \\ 3 & -1 \\ 4 & 1 \end{pmatrix} \begin{pmatrix} x_1 \\ x_2 \end{pmatrix} = \begin{pmatrix} 3 \\ 1 \\ -1 \end{pmatrix}$$

$$AX = B$$

$$A = \begin{pmatrix} 2 & 5 \\ 3 & -1 \\ 4 & 1 \end{pmatrix}; \ [A/B] = \begin{pmatrix} 2 & 5 & 3 \\ 3 & -1 & 1 \\ 4 & 1 & -1 \end{pmatrix}$$

In this case, we need to first determine $rank([A/B])$. This is because the greatest possible rank of A is 2; that is, the largest square submatrix of A is of order 2x2, but $rank([A/B])$ could be 3.

If $rank([A/B]) = 3$, then

$$rank(A) < rank([A/B]).$$

If $rank([A/B]) \neq 3$, and $rank = 2$, then

$$rank(A) = rank([A/B]).$$

It may well happen that $rank([A/B]) = rank\ (A) < 2$, the largest possible rank of A, in which case the system has an infinite number of solutions. Let's go.

For $rank\ ([A/B])$, we have

$$\begin{pmatrix} 2 & 5 & 3 \\ 3 & -1 & 1 \\ 4 & 1 & -1 \end{pmatrix} (R_3 - 2R_1) \sim \begin{pmatrix} 2 & 5 & 3 \\ 3 & -1 & 1 \\ 0 & -9 & -7 \end{pmatrix} (3R_1),(2R_2)(R_2 - R_1) \sim \begin{pmatrix} 6 & 15 & 9 \\ 0 & -17 & -7 \\ 0 & -9 & -7 \end{pmatrix}$$

$$(R_3 - \tfrac{19}{17} R_2),\ (17R_3) \sim \begin{pmatrix} 6 & 15 & 9 \\ 0 & -17 & -7 \\ 0 & 0 & -182 \end{pmatrix}$$

$\Rightarrow\ rank([A\ /\ B]) = 3 > rank\ (A)$. The system has no solution.

(b) $\quad x_1 - x_2 = 3$

$\qquad 2x_1 + 3x_2 = 5$

$\qquad x_1 - 6x_2 = 4$

$$[A/\ B] = \begin{pmatrix} 1 & -1 & 3 \\ 2 & 3 & 5 \\ 1 & -6 & 4 \end{pmatrix} (R_3 - (3R_1 - R_2)) \sim \begin{pmatrix} 1 & -1 & 3 \\ 2 & 3 & 5 \\ 0 & 0 & 0 \end{pmatrix} (R_2 - 2R_1) \sim \begin{pmatrix} 1 & -1 & 3 \\ 0 & 5 & -1 \\ 0 & 0 & 0 \end{pmatrix}$$

$$\Rightarrow rank\ ([A/B]) = 2$$

Since $rank([A\ /\ B]) = rank(A) = 2$, the system has a unique solution.

This implies, in this case, that anyone of the equations can be written as a linear combination of the other two. Instead of reducing to the echelon form, we could have gone ahead to just test whether the determinant of $[A/B]$ is zero or not.

$$\begin{vmatrix} 1 & -1 & 3 \\ 2 & 3 & 5 \\ 1 & -6 & 4 \end{vmatrix} = 0$$

This implies that the system has a solution; that is, the system is consistent. This determinant method, however, does not tell us whether the solution is unique or not. This is because it only tells us that $rank([A/B]) < 3$, which could be 1 or 2. Therefore, we also need to further consider the 2x2 submatrices of A. If at least one 2x2 submatrix of A is nonsingular, then we are certain that $rank([A/B]) = 2$. If all the possible 2x2 submatrices of A are singular, then $rank(A) = 1 = rank([A/B]) < 2$, which means that the system has infinitely many solutions. The echelon method is more specific.

c. $3x_1 + x_2 - x_3 = 0$

$-x_1 + 2x_2 + x_3 = 0$

$2x_1 - x_2 + 2x_3 = 0$

$$A = \begin{pmatrix} 3 & 1 & -1 \\ -1 & 2 & 1 \\ 2 & -1 & 2 \end{pmatrix} \quad [A/B] = \begin{pmatrix} 3 & 1 & -1 & 0 \\ -1 & 2 & 1 & 0 \\ 2 & -1 & 2 & 0 \end{pmatrix}$$

Without performing any calculation, it is obvious that the rank of A must be equal to that of $[A/B]$ due to the zero column of $[A/B]$. Recall that the determinant of a matrix having a row or a column of zeros is equal to zero. This means that the determinant of any 3x3 submatrix of $[A/B]$ that includes the zero- column is zero. This further implies that the largest nonsingular submatrix of $[A/B]$ must be equal to that of the matrix A part of $[A/B]$ – meaning that $rank(A) = rank([A/B])$.

This is not an accident. It is always the case that for any linear system of the form

$$AX = 0$$

$$rank(A) = rank([A/B]).$$

This means that the system $AX = 0$ is always consistent; that is, it always has a solution-at least, the *trivial solution*

$$\begin{pmatrix} x_1 \\ x_2 \\ \vdots \\ \vdots \\ x_n \end{pmatrix} = \begin{pmatrix} 0 \\ 0 \\ \vdots \\ \vdots \\ 0 \end{pmatrix}$$

The system $AX = 0$ is called *homogenous system*. It always has at least one solution. It is easy to see that the null vector

$$\begin{pmatrix} 0 \\ 0 \\ \vdots \\ \vdots \\ 0 \end{pmatrix}$$ is always a solution of a homogeneous system. If this *trivial* solution is the

only solution, that means the system is unique, and thus

$$rank(A) = rank([A/B]) = n \Rightarrow |A| \neq 0$$

If the system is not unique; it has infinitely many solutions; thus

$$rank(A) = rank([A/B]) < n \Rightarrow |A| = 0$$

In summary, the homogenous system $AX = 0$

i)   *has exactly one solution, which is the trivial solution if and only if*

$$|A| \neq 0$$

ii)   *has infinitely many solutions if and only if*

$$|A| = 0$$

Example 1.10:

Find the value of t such that the system

$2x_1 + 3x_2 - x_3 = 0$

$-x_1 + x_3 = 0$

$3x_1 + tx_2 = 0$

has an infinite number of solutions.

Solution

The system is homogenous, and so for it to have more than one solution (infinite number of solutions) it is required that

$|A| = 0$, where A is the coefficient matrix, we must have

$$\begin{vmatrix} 2 & 3 & -1 \\ -1 & 0 & 1 \\ 3 & t & 0 \end{vmatrix} = 0$$

$$\Rightarrow \begin{vmatrix} 1 & 3 & 0 \\ -1 & 0 & 1 \\ 3 & t & 0 \end{vmatrix} = 0$$

Expanding along the third column, we have

$$(-1)^{2+3}(1)(t - 9) = 0$$

$$\therefore t = 9$$

[Note: we simplified the determinant before expanding. This was achieved by replacing row 1 with its sum with row 2, which does not alter the determinant.]

Again, note that there is no value of t for which the system has no solution. This is because $rank(A)$ is always equal to $rank([A/B])$ – meaning that it will always have a solution, at worst the trivial solution.

Linear dependence and linear independence

A collection of vectors $u_1, u_2, \dots u_n$ in a *vector space* V over a scalar field K is said to be linearly dependent if there exist scalars a_i $(1 \leq i \leq n)$ in K not all zero such that

$a_1 u_1 + a_2 u_2 + \cdots a_n u_n = 0$, otherwise the collection of vectors is said to be linearly independent. In other words, the set of vectors $u_1, u_2, \dots u_n$ in V is said to be linearly independent if the equation

$a_1 u_1 + a_2 u_2 + \cdots a_n u_n = 0$ *always implies that* $a_1 = a_2 = \cdots a_n = 0$.

In summary, if S is a subset of a vector space V, over a scalar field K, then $S = \{u_1, u_2, \ldots u_n\}$ is linearly dependent if

$$\sum_{i=1}^{n} a_i u_i = 0$$

$\Rightarrow$ at least one $a_i \neq 0$; $a_i \in K$ for $each$ $i \in \{1,2 \ldots n\}$.

The set $S = \{u_1, u_2 \ldots u_n\}$ is linearly independent if

$$\sum_{i=1}^{n} a_i u_i = 0$$

$\Rightarrow a_i = 0$; $a_i \in K, for$ $each$ $i \in \{1,2 \ldots n\}$.

Let

$$u_i = \begin{pmatrix} u_{i1} \\ u_{i2} \\ u_{i3} \\ \vdots \\ \vdots \\ u_{ik} \end{pmatrix} for$$ $each$ $i \in \{1,2 \ldots n\}$, then

$$\sum_{i=1}^{n} a_i u_i = 0 \Rightarrow$$

$$
\begin{array}{lllll}
a_1 u_{11} + & a_2 u_{21} + & a_3 u_{31} + & \ldots & a_n u_{n1} = 0 \\
a_1 u_{12} + & a_2 u_{22} + & a_3 u_{32} + \cdots & & a_n u_{n2} = 0 \\
\vdots & \vdots & \vdots & \vdots & \ldots\ldots\ldots\ldots@ \\
\vdots & \vdots & \vdots & \vdots & \\
a_1 u_{1k} + & a_2 u_{2k} + & a_3 u_{3k} + & & a_n u_{nk} = 0
\end{array}
$$

The system @ is a **homogenous** linear system of k equations in n unknowns, which can be represented in matrix form as

$$UA = \mathbf{0}$$

where

$$U = \begin{pmatrix} u_{11} & u_{21} \ldots \ldots \ldots & & u_{n1} \\ u_{12} & u_{22} \ldots \ldots \ldots & & \\ \vdots & \vdots & & \\ \vdots & \vdots & & \\ u_{1k} & u_{2k} \ldots \ldots \ldots \ldots & & u_{nk} \end{pmatrix} \quad ; \quad A = \begin{pmatrix} a_1 \\ a_2 \\ \vdots \\ \vdots \\ a_n \end{pmatrix}$$

Case 1: $k = n$

Recall that for a homogeneous linear system, the system assumes only the trivial solution, $a_1 = a_2 = \ldots \ldots a_n = 0$ if and only if $|U| \neq 0$. Note, also that $|U| = 0$ if and only if $|U^T| = 0$, a property of determinant. We deal with U^T for pedagogical reasons.

Therefore, for linear independence,

$$|U^T| = \begin{vmatrix} u_{11} & u_{12} \ldots & u_{1n} \\ u_{21} & u_{22} \ldots & u_{2n} \\ \vdots & \vdots & \vdots \\ \vdots & \vdots & \vdots \\ u_{n1} & u_{n2} \ldots & u_{nn} \end{vmatrix} \neq 0 \qquad\qquad ----- P161$$

If at least one a_i is not equal to zero, then the system @ must have more than one solution, hence an infinite number of solutions; thus,

$$|U^T| = \begin{vmatrix} u_{11} & u_{12} \ldots & u_{1n} \\ u_{21} & u_{22} \ldots & u_{2n} \\ \vdots & \vdots & \vdots \\ \vdots & \vdots & \vdots \\ u_{n1} & u_{n2} \ldots & u_{nn} \end{vmatrix} = 0 \qquad\qquad ----- P162$$

Equation $P161$ and $P162$ are the conditions for linear independence and linear dependence respectively.

Case 2: $k > n$

In this case there are more equations than unknowns, and so it becomes meaningless to talk of the determinant of the coefficient matrix, being a non-square matrix. We therefore resort to the more general way of testing for linear dependence and linear independence, the row-echelon reduction. In this case, and even in case 1, we place each k-dimensional vector as a row vector so as to form an $n \times k$ matrix, which, in this case, is not a square matrix. Whether a square matrix or not, we perform the elementary row operations on the $n \times k$ coefficient matrix until it is reduced to an echelon matrix; if there is at least one zero row, then the set of vectors is linearly dependent, otherwise the set is linearly independent.

Case 3: $k < n$

In this case, there are more unknowns than equations. Since the system, being homogeneous, has at least one solution, therefore the system either has only the trivial solution or infinitely many solutions. However, the fact that any linear system with more unknowns than equations either has no solution (inconsistent) or has an infinite number of solutions leaves us with only one option: the case $k < n$ has an infinite number of solutions and so there must be scalars a_i not all zero such that

$$\sum_{i=1}^{n} a_i \boldsymbol{u}_i = \boldsymbol{0}.$$

Therefore this case is always linearly dependent.

To see that a linear system with more unknowns than equations cannot be unique, consider the system below:

$$a_{11}x_1 + a_{12}x_2 + a_{1n}x_n = b_1$$
$$a_{21}x_1 + a_{22}x_2 \quad a_{2n}x_n = b_2$$
$$\vdots \qquad \vdots \qquad \vdots$$
$$\vdots \qquad \vdots \qquad \vdots$$
$$a_{k1}x_1 + a_{k2}x_2 \quad \ldots a_{kn}x_n = b_k$$

$$\begin{pmatrix} a_{11} & a_{12} & \ldots & a_{1n} \\ a_{21} & a_{22} & \ldots & a_{2n} \\ \vdots & \vdots & \vdots & \vdots \\ \vdots & \vdots & \vdots & \vdots \\ a_{k1} & a_{k2} & \ldots & a_{kn} \end{pmatrix} \begin{pmatrix} x_1 \\ x_2 \\ \vdots \\ \vdots \\ x_n \end{pmatrix} = \begin{pmatrix} b_1 \\ b_2 \\ \vdots \\ \vdots \\ b_k \end{pmatrix}$$

In this system, we assume that the number of equations is less than that of unknowns; that is, $k < n$.

If A is the coefficient matrix, then $rank(A)$ is at most k, since k is less than n; therefore, $rank(A) < n$, the number of unknowns – meaning that the system is either inconsistent or nonunique.

Example 1.11:

Determine if the following set of vectors in R^3 is linearly dependent or not.

$$\{v_1 = (2, -3, 1), v_2 = (3, 1, 2), v_3 = (-1, 1, 0)\}$$

Solution

In this case, $n = k$. Let us use the determinant method, since it is easier to compute the determinant of just one 3x3 matrix than reducing it to its row – echelon form. Placing the vectors as rows, and finding the determinant, we have

$$D = \begin{vmatrix} 2 & -3 & 1 \\ 3 & 1 & 2 \\ -1 & 1 & 0 \end{vmatrix}$$

Expanding along the third row, we have

$$D = (-1)^{3+1}(-1)(2(-3) - 1) + (-1)^{3+2}(1)(2(2) - 3(1)) + (-1)^{3+3}(0)(\dots)$$

$$D = 6 \neq 0.$$

Therefore the set is linearly independent.

Example 1.12:

Determine whether the following sets are dependent or not.

a) $\{(3, -1), (4, 2), (-1, 5)\}$ in R^2

b) $\{(2, 1, 3, 1), (1, 0, 1, -1), (0, -1, 1, -3)\}$ in R^4

Solution

(a) We do not need to perform any calculations. This case corresponds to the case 3 above, which is always linearly dependent. This is because solving the equation

$\mathbf{0} = a_1(3, -1) + a_2(4, 2) + a_3(-1, 5)$ will always results in finding the unknowns $a_1, a_2,$ and a_3 in two consistent equations. This ensures that at least one of them is not equal to zero. Therefore the set of vectors is linearly dependent.

In general, any set of n distinct k – dimensional vectors with $k < n$, is always linearly dependent. That is, any set of three or more distinct 2– dimensional vectors is linearly dependent; any set of four or more distinct 3– dimensional vectors is linearly dependent, and so on.

(b) To ascertain this set, we need to row-reduce the matrix

$$\begin{pmatrix} 2 & 1 & 3 & 1 \\ 1 & 0 & 1 & -1 \\ 0 & -1 & 1 & -3 \end{pmatrix}.$$

$$\begin{pmatrix} 2 & 1 & 3 & 1 \\ 1 & 0 & 1 & -1 \\ 0 & -1 & 1 & -3 \end{pmatrix} (R_1 \sim R_2), (R_2 - 2R_1) \begin{pmatrix} 1 & 0 & 1 & -1 \\ 0 & 0 & 1 & 3 \\ 0 & -1 & 1 & -3 \end{pmatrix}$$

$$(R_3 + R_2) \begin{pmatrix} 1 & 0 & 1 & -1 \\ 0 & 1 & 1 & -3 \\ 0 & 0 & 2 & 0 \end{pmatrix}$$

Since there is no zero-row in this echelon form, therefore the set is linearly independent.

Note that, although the determinant of a nonsquare matrix is not defined, we could have used the determinant method by evaluating the determinant of the largest square submatrices. If at least one of such matrices is nonsingular, then the set is linearly independent, otherwise it is linearly dependent.

In general, testing if a set of n vectors from the same vector space is linearly dependent or not is equivalent to determining the rank of the matrix formed by the vectors (placed either as columns or as rows). If the rank is less than n, then the system is linearly dependent; if the rank is equal to n, then the system is linearly independent. The rank cannot be greater than n.

SOLUTION METHODS

We dedicate the last few pages of this chapter to different solution methods to a system of linear algebraic equations.

THE *LU* Decomposition Method

LU is a contraction of "lower triangular, upper triangular" that is, lower triangular matrix and upper triangular matrix. This method employs the fact that a square matrix A can be written as a product of a lower triangular matrix and an upper triangular matrix provided that matrix A is nonsingular.

Let A, a square matrix, be written as

$$A = LU$$

Where

$$L = \begin{pmatrix} l_{11} & 0 & 0 \\ l_{21} & l_{22} & 0 \\ l_{31} & l_{32} & l_{33} \end{pmatrix}, U = \begin{pmatrix} u_{11} & u_{12} & u_{13} \\ 0 & u_{22} & u_{23} \\ 0 & 0 & u_{33} \end{pmatrix}$$

$$A = \begin{pmatrix} l_{11} & 0 & 0 \\ l_{21} & l_{22} & 0 \\ l_{31} & l_{32} & l_{33} \end{pmatrix}\begin{pmatrix} u_{11} & u_{12} & u_{13} \\ 0 & u_{22} & u_{23} \\ 0 & 0 & u_{33} \end{pmatrix}$$

$$A = \begin{pmatrix} l_{11}u_{11} & l_{11}u_{12} & l_{11}u_{13} \\ l_{21}u_{11} & l_{21}u_{12}+l_{22}u_{22} & l_{21}u_{13}+l_{22}u_{23} \\ l_{31}u_{11} & l_{31}u_{12}+l_{32}u_{22} & l_{31}u_{13}+l_{32}u_{23} + l_{33}u_{33} \end{pmatrix}$$

Assuming $u_{11} = u_{22} = u_{33} = 1$, then

$$A = \begin{pmatrix} l_{11} & l_{11}u_{12} & l_{11}u_{13} \\ l_{21} & l_{21}u_{12}+l_{22} & l_{21}u_{13}+l_{22}u_{23} \\ l_{31} & l_{31}u_{12}+l_{32} & l_{31}u_{13}+l_{32}u_{23} + l_{33} \end{pmatrix}$$

Note that there is no harm in the assumption that $u_{11} = u_{22} = u_{33} = 1$. For if $u_{11} \neq 1$, then we can divide all the elements of U by u_{11} so that matrix A becomes

$$A = \begin{pmatrix} l_{11} & 0 & 0 \\ l_{21} & l_{22} & 0 \\ l_{31} & l_{32} & l_{33} \end{pmatrix} u_{11} \begin{pmatrix} 1 & \dfrac{u_{12}}{u_{11}} & \dfrac{u_{13}}{u_{11}} \\ 0 & \dfrac{u_{22}}{u_{11}} & \dfrac{u_{23}}{u_{11}} \\ 0 & 0 & \dfrac{u_{33}}{u_{11}} \end{pmatrix}$$

$$A = \begin{pmatrix} l_{11}u_{11} & 0 & 0 \\ l_{21}u_{11} & l_{22}u_{11} & 0 \\ l_{31}u_{11} & l_{32}u_{11} & l_{33}u_{11} \end{pmatrix} \begin{pmatrix} 1 & \dfrac{u_{12}}{u_{11}} & \dfrac{u_{13}}{u_{11}} \\ 0 & \dfrac{u_{22}}{u_{11}} & \dfrac{u_{23}}{u_{11}} \\ 0 & 0 & \dfrac{u_{33}}{u_{11}} \end{pmatrix}$$

$$\Rightarrow \quad A = \begin{pmatrix} p_{11} & 0 & 0 \\ p_{21} & p_{22} & 0 \\ p_{31} & p_{32} & p_{33} \end{pmatrix} \begin{pmatrix} 1 & q_{12} & q_{13} \\ 0 & q_{22} & q_{23} \\ 0 & 0 & q_{33} \end{pmatrix}, \text{ which is obviously in}$$

LU form with $u_{11} = 1$. Therefore, A can always be written as LU with $u_{11} = 1$.

$$\Rightarrow A = \begin{pmatrix} l_{11} & l_{11}u_{12} & l_{11}u_{13} \\ l_{21} & l_{21}u_{12} + l_{22}u_{22} & l_{21}u_{13} + l_{22}u_{23} \\ l_{31} & l_{31}u_{12} + l_{32}u_{22} & l_{31}u_{13} + l_{32}u_{23} + l_{33}u_{33} \end{pmatrix} \text{ is justified.}$$

Again, note that u_{22} only appears in the terms $l_{22}u_{22}$ and $l_{32}u_{22}$ in the product matrix; therefore, we can set $u_{22} = 1$, and then set both l_{22} and l_{32} in L as $l_{22}u_{22}$ and $l_{32}u_{22}$ respectively. Similarly, we set $u_{33} = 1$, and then set l_{33} in L as $l_{33}u_{33}$ since u_{33} appears only in the term $l_{33}u_{33}$ in the product matrix. Therefore,

$A = LU$ can be written as

$$A = \begin{pmatrix} l_{11} & 0 & 0 \\ l_{21} & l_{22} & 0 \\ l_{31} & l_{32} & l_{33} \end{pmatrix} \begin{pmatrix} u_{11} & u_{12} & u_{13} \\ 0 & u_{22} & u_{23} \\ 0 & 0 & u_{33} \end{pmatrix} = \begin{pmatrix} l_{11} & l_{11}u_{12} & l_{11}u_{13} \\ l_{21} & l_{21}u_{12} + l_{22} & l_{21}u_{13} + l_{22}u_{23} \\ l_{31} & l_{31}u_{12} + l_{32} & l_{31}u_{13} + l_{32}u_{23} + l_{33} \end{pmatrix}$$

Example 1.13

Write the matrix $A = \begin{pmatrix} 3 & -1 & 1 \\ 1 & 1 & -1 \\ 1 & -2 & 3 \end{pmatrix}$ in LU form; hence solve the system

$3x_1 - x_2 + x_3 = 4$

$x_1 + x_2 - 3x_3 = 4$

$x_1 - 2x_2 + 3x_3 = -5$

Solution

We simply need to compare matrix A with the standard $LU -$ form matrix already established. That is,

$$\begin{pmatrix} 3 & -1 & 1 \\ 1 & 1 & -1 \\ 1 & -2 & 3 \end{pmatrix} = \begin{pmatrix} l_{11} & l_{11}u_{12} & l_{11}u_{13} \\ l_{21} & l_{21}u_{12} + l_{22} & l_{21}u_{13} + l_{22}u_{23} \\ l_{31} & l_{31}u_{12} + l_{32} & l_{31}u_{13} + l_{32}u_{23} + l_{33} \end{pmatrix}$$

$$\Rightarrow \quad l_{11} = 3, l_{21} = 1, l_{31} = 1$$

$$l_{11}u_{12} = -1, l_{11}u_{13} = 1$$

$$u_{12} = \frac{-1}{3}, u_{13} = \frac{1}{3}$$

$$l_{21}u_{12} + l_{22} = 1 \quad, \quad l_{31}u_{12} + l_{32} = -2$$

$$(1)\left(\frac{-1}{3}\right) + l_{22} = 1, (1)\left(\frac{-1}{3}\right) + l_{32} = -2$$

$$l_{22} = \frac{4}{3}, l_{32} = \frac{-5}{3}$$

Finally,

$$l_{21}u_{13} + l_{22}u_{23} = -1 \quad, \quad l_{31}u_{13} + l_{32}u_{23} + l_{33} = 3$$

$$(1)\left(\frac{1}{3}\right) + \frac{4}{3}u_{23} = -1, (1)\left(\frac{1}{3}\right) + \frac{-5}{3}u_{23} + l_{33} = 3$$

$$u_{23} = -1, l_{33} = 1$$

$$A = \begin{pmatrix} l_{11} & 0 & 0 \\ l_{21} & l_{22} & 0 \\ l_{31} & l_{32} & l_{33} \end{pmatrix} \begin{pmatrix} u_{11} & u_{12} & u_{13} \\ 0 & u_{22} & u_{23} \\ 0 & 0 & u_{33} \end{pmatrix}$$

$$A = \begin{pmatrix} 3 & 0 & 0 \\ 1 & \frac{4}{3} & 0 \\ 1 & \frac{-5}{3} & 1 \end{pmatrix} \begin{pmatrix} 1 & \frac{-1}{3} & \frac{1}{3} \\ 0 & 1 & -1 \\ 0 & 0 & 1 \end{pmatrix}$$

Now let us use our result to solve the system

$$AX = B$$

Since

$$A = LU$$

$$LUX = B$$

$$L(UX) = B$$

The matrix UX is a 3 x 1 matrix because $U = 3 \times 3$, and $X = 3 \times 1$

Let

$$UX = \begin{pmatrix} y_1 \\ y_2 \\ y_3 \end{pmatrix} = Y$$

$$LY = B$$

$$\Rightarrow \begin{pmatrix} 3 & 0 & 0 \\ 1 & \dfrac{4}{3} & 0 \\ 1 & \dfrac{-5}{3} & 1 \end{pmatrix} \begin{pmatrix} y_1 \\ y_2 \\ y_3 \end{pmatrix} = \begin{pmatrix} 4 \\ 4 \\ -5 \end{pmatrix}$$

$$3y_1 = 4 \,, y_1 = \frac{4}{3}$$

$$y_1 + \frac{4}{3}y_2 = 4$$

$$\frac{4}{3} + \frac{4}{3}y_2 = 4 \,, y_2 = 2$$

Again

$$y_1 - \frac{5}{3}y_2 + y_3 = -5 \,, \frac{4}{3} - \frac{5}{3}(2) + y_3 = -5 \,, \ y_3 = -3$$

But

$$y = \begin{pmatrix} y_1 \\ y_2 \\ y_3 \end{pmatrix} = UX$$

$$\begin{pmatrix} \frac{4}{3} \\ 2 \\ -3 \end{pmatrix} = \begin{pmatrix} 1 & \dfrac{-1}{3} & \dfrac{1}{3} \\ 0 & 1 & -1 \\ 0 & 0 & 1 \end{pmatrix} \begin{pmatrix} x_1 \\ x_2 \\ x_3 \end{pmatrix}$$

$$\Rightarrow x_3 = -3 \,, x_2 - x_3 = 2 \,, x_2 = -1$$

$$x_1 - \frac{1}{3}x_2 + \frac{1}{3}x_3 = \frac{4}{3} \ , \ \ x_1 = 2$$

$$\Rightarrow \begin{pmatrix} x_1 \\ x_2 \\ x_3 \end{pmatrix} = \begin{pmatrix} 2 \\ -1 \\ -3 \end{pmatrix}$$

Example 1.14:

Solve the following linear system, using the *LU* method.

$3x_1 - x_2 + 2x_3 = -1$

$-x_1 + 5x_2 + 10x_3 = 13$

$2x_1 - 3x_2 - 8x_3 = -3$

Solution

As before,

$$A = \begin{pmatrix} 3 & -1 & 2 \\ -1 & 5 & 10 \\ 2 & -3 & -8 \end{pmatrix} = \begin{pmatrix} l_{11} & l_{11}u_{12} & l_{11}u_{13} \\ l_{21} & l_{21}u_{12} + l_{22} & l_{21}u_{13} + l_{22}u_{23} \\ l_{31} & l_{31}u_{12} + l_{32} & l_{31}u_{13} + l_{32}u_{23} + l_{33} \end{pmatrix}$$

$$\Rightarrow \quad l_{11} = 3 \,, l_{21} = -1 \,, l_{31} = 2$$

$$l_{11}u_{12} = -1 \,, l_{11}u_{13} = 2$$

$$u_{12} = \frac{-1}{3} \,, u_{13} = \frac{2}{3}$$

$$l_{21}u_{12} + l_{22} = 5 \ , \qquad l_{31}u_{12} + l_{32} = -3$$

$$l_{22} = \frac{14}{3} \quad , \qquad l_{32} = \frac{-7}{3}$$

Finally,

$$l_{21}u_{13} + l_{22}u_{23} = 10 \ , \qquad l_{31}u_{13} + l_{32}u_{23} + l_{33} = -8$$

$$u_{23} = \frac{16}{7} \quad , \qquad l_{33} = -4$$

$$A = \begin{pmatrix} l_{11} & 0 & 0 \\ l_{21} & l_{22} & 0 \\ l_{31} & l_{32} & l_{33} \end{pmatrix} \begin{pmatrix} 1 & u_{12} & u_{13} \\ 0 & 1 & u_{23} \\ 0 & 0 & 1 \end{pmatrix}$$

$$A = \begin{pmatrix} 3 & 0 & 0 \\ -1 & \dfrac{14}{3} & 0 \\ 2 & \dfrac{-7}{3} & -4 \end{pmatrix} \begin{pmatrix} 1 & \dfrac{-1}{3} & \dfrac{2}{3} \\ 0 & 1 & \dfrac{16}{7} \\ 0 & 0 & 1 \end{pmatrix}$$

$$AX = B$$
$$LUX = B$$
$$L(UX) = B$$

Let

$$UX = Y = \begin{pmatrix} y_1 \\ y_2 \\ y_3 \end{pmatrix}$$

$$LY = B$$

$$\begin{pmatrix} 3 & 0 & 0 \\ -1 & \frac{14}{3} & 0 \\ 2 & \frac{-7}{3} & -4 \end{pmatrix} \begin{pmatrix} y_1 \\ y_2 \\ y_3 \end{pmatrix} = \begin{pmatrix} -1 \\ 13 \\ -3 \end{pmatrix}$$

$$3y_1 = -1 \, , y_1 = \frac{-1}{3}$$

$$-y_1 + \frac{14}{3} y_2 = 13$$

$$\frac{4}{3} + \frac{14}{3} y_2 = 13 \, , \; y_2 = \frac{19}{7}$$

Again,

$$2y_1 - \frac{7}{3} y_2 - 4y_3 = -3 \, , \; y_3 = -1$$

$$\begin{pmatrix} \frac{-1}{3} \\ \frac{19}{7} \\ -1 \end{pmatrix} = \begin{pmatrix} 1 & \frac{-1}{3} & \frac{2}{3} \\ 0 & 1 & \frac{16}{7} \\ 0 & 0 & 1 \end{pmatrix} \begin{pmatrix} x_1 \\ x_2 \\ x_3 \end{pmatrix}$$

$$\Rightarrow x_3 = -1$$

$$x_2 + \frac{16}{7}x_3 = \frac{19}{7} \ , \ x_2 = 5$$

$$x_1 - \frac{1}{3}x_2 + \frac{2}{3}x_3 = \frac{-1}{3} \ , \ x_1 = 2$$

$$\Rightarrow \begin{pmatrix} x_1 \\ x_2 \\ x_3 \end{pmatrix} = \begin{pmatrix} 2 \\ 5 \\ -1 \end{pmatrix}$$

The Inverse Method

This is another method for solving the linear system

$$AX = B$$

Let us left multiply both sides of the equation by A^{-1}, the inverse of A, keeping in mind that $A^{-1}A = I$, we have

$$A^{-1}AX = A^{-1}B; \text{ provided } |A| \neq 0$$

$$\Rightarrow \quad IX = A^{-1}B$$

$$X = A^{-1}B, \text{ the solution vector.}$$

[**Note:** left multiplying or right multiplying both sides of a matrix equation by a matrix other than a null matrix does not alter the equation]

Example 1.15:

Solve the following linear system, using the inverse method.

$$2x_1 - 3x_2 = 8$$
$$3x_1 + 5x_2 = 31$$

Solution

We first write the system in matrix form:

$$\begin{pmatrix} 2 & -3 \\ 3 & 5 \end{pmatrix} \begin{pmatrix} x_1 \\ x_2 \end{pmatrix} = \begin{pmatrix} 8 \\ 31 \end{pmatrix}$$

We need to find the inverse of the matrix $A = \begin{pmatrix} 2 & -3 \\ 33 & 5 \end{pmatrix}$. Recall that inverse of a matrix is defined by

$$AA^{-1} = A^{-1}A = I$$

Let

$$A^{-1} = \begin{pmatrix} u_{11} & u_{12} \\ u_{21} & u_{22} \end{pmatrix}$$

$$\begin{pmatrix} u_{11} & u_{12} \\ u_{21} & u_{22} \end{pmatrix} \begin{pmatrix} 2 & -3 \\ 3 & 5 \end{pmatrix} = \begin{pmatrix} 1 & 0 \\ 0 & 1 \end{pmatrix}$$

$$\begin{pmatrix} 2u_{11} + 3u_{12} & -3u_{11} + 5u_{12} \\ 2u_{21} + 3u_{22} & -3u_{21} + 5u_{22} \end{pmatrix} = \begin{pmatrix} 1 & 0 \\ 0 & 1 \end{pmatrix}$$

$$2u_{11} + 3u_{12} = 1 \quad , \quad -3u_{11} + 5u_{12} = 0$$

$$2u_{21} + 3u_{22} = 0 \quad , \quad -3u_{21} + 5u_{22} = 1$$

$$\Rightarrow u_{11} = \frac{5}{19}, \quad u_{12} = \frac{3}{19}, \quad u_{21} = \frac{-3}{19}, \quad u_{22} = \frac{2}{19}$$

$$A^{-1} = \begin{pmatrix} \dfrac{5}{19} & \dfrac{3}{19} \\ \dfrac{-3}{19} & \dfrac{2}{19} \end{pmatrix} = \frac{1}{19} \begin{pmatrix} 5 & 3 \\ -3 & 2 \end{pmatrix}$$

$$x = \begin{pmatrix} x_1 \\ x_2 \end{pmatrix} = A^{-1}B$$

$$\Rightarrow \begin{pmatrix} x_1 \\ x_2 \end{pmatrix} = \frac{1}{19} \begin{pmatrix} 5 & 3 \\ -3 & 2 \end{pmatrix} \begin{pmatrix} 8 \\ 31 \end{pmatrix}$$

$$\begin{pmatrix} x_1 \\ x_2 \end{pmatrix} = \frac{1}{19}\begin{pmatrix} 40 & + & 93 \\ -24 & + & 62 \end{pmatrix} = \begin{pmatrix} 7 \\ 2 \end{pmatrix}, \text{ the solution vector.}$$

The above method of finding inverse is not efficient, especially when the order of the matrix is substantially high. To find the inverse of A in a more efficient way, we simply write the concatenated matrix

$$(A/I) \ or \ [A/B]$$

and then perform elementary row operations on it until the matrix A part turns to the identity matrix I. Having done this, the matrix I becomes A^{-1}. This method is called the **row-transformation** method. Let us use it to find the inverse of matrix A above.

$$A = \begin{pmatrix} 2 & -3 \\ 3 & 5 \end{pmatrix}$$

Using the method, we have

$$\begin{pmatrix} 2 & -3 & 1 & 0 \\ 3 & 5 & 0 & 1 \end{pmatrix} (\tfrac{1}{2} R_1), (R_2 - 3R_1) \begin{pmatrix} 1 & \frac{-3}{2} & \frac{1}{2} & 0 \\ 0 & \frac{19}{2} & \frac{-3}{2} & 1 \end{pmatrix}$$

$$(\tfrac{2}{19}R_2) \begin{pmatrix} 1 & \frac{-3}{2} & \frac{1}{2} & 0 \\ 0 & 1 & \frac{-3}{19} & \frac{2}{19} \end{pmatrix} (R_1 + \tfrac{3}{2}R_2) \begin{pmatrix} 1 & 0 & \frac{5}{19} & \frac{3}{19} \\ 0 & 1 & \frac{-3}{19} & \frac{2}{19} \end{pmatrix}$$

Therefore, the inverse of $\begin{pmatrix} 2 & -3 \\ 3 & 5 \end{pmatrix}$ is $\begin{pmatrix} \frac{5}{19} & \frac{3}{19} \\ \frac{-3}{19} & \frac{2}{19} \end{pmatrix}$.

We then use the inverse to left multiply the vector B, the right hand side of the system, to obtain the solution vector.

Example 1.16:

Find the inverse of matrix $A = \begin{pmatrix} a & c \\ b & d \end{pmatrix}$.

Solution

$$\left(\begin{array}{cc|cc} a & c & 1 & 0 \\ b & d & 0 & 1 \end{array}\right) \underset{\sim}{\tfrac{1}{a}R_1} \left(\begin{array}{cc|cc} 1 & \tfrac{c}{a} & \tfrac{1}{a} & 0 \\ b & d & 0 & 1 \end{array}\right) \underset{\sim}{(R_2 - bR_1)}$$

$$\left(\begin{array}{cc|cc} 1 & \tfrac{c}{a} & \tfrac{1}{a} & 0 \\ 0 & d-\tfrac{bc}{a} & \tfrac{-b}{a} & 1 \end{array}\right) \underset{\sim}{\overset{aR_2}{\scriptstyle ad-bc}} \left(\begin{array}{cc|cc} 1 & \tfrac{c}{a} & \tfrac{1}{a} & 0 \\ 0 & 1 & \tfrac{-b}{ad-bc} & \tfrac{a}{ad-bc} \end{array}\right)$$

$$\underset{\sim}{(R_1 - \tfrac{c}{a}R_2)} \left(\begin{array}{cc|cc} 1 & 0 & \tfrac{d}{ad-bc} & \tfrac{-c}{ad-bc} \\ 0 & 1 & \tfrac{-b}{ad-bc} & \tfrac{a}{ad-bc} \end{array}\right)$$

$$\begin{pmatrix} a & c \\ b & d \end{pmatrix}^{-1} = \begin{pmatrix} \dfrac{d}{ad-bc} & \dfrac{-c}{ad-bc} \\ \dfrac{-b}{ad-bc} & \dfrac{a}{ad-bc} \end{pmatrix} = A^{-1}$$

$$A^{-1} = \frac{1}{ad-bc}\begin{pmatrix} d & -c \\ -b & a \end{pmatrix} \qquad\qquad -----P163$$

Notice that all the entries (elements) of A^{-1} naturally have a common denominator, $ad - bc = |A|$; also, the diagonal elements of A are interchanged, and the signs of the other two elements are changed (multiplied by -1). Since A is a general 2 x 2 matrix, it is always the case that the inverse of any nonsingular 2x2 matrix is obtained by interchanging the main diagonal elements, changing the signs of the other two entries, and then dividing the new matrix by the determinant of the original matrix.

[Note: dividing a matrix by a scalar means multiplying the matrix by the reciprocal of the scalar. It also means dividing each entry by the scalar.]

$$\text{If } A = \begin{pmatrix} a & c \\ b & d \end{pmatrix}, \text{then } A^{-1} = \frac{1}{|A|} \begin{pmatrix} d & -c \\ -b & a \end{pmatrix}$$

For instance, the inverse of $A = \begin{pmatrix} a & -b \\ b & a \end{pmatrix}$ is

$$A^{-1} = \frac{\begin{pmatrix} a & b \\ -b & a \end{pmatrix}}{a^2 + b^2} = \frac{1}{a^2 + b^2} \begin{pmatrix} a & b \\ -b & a \end{pmatrix} = \begin{pmatrix} \dfrac{a}{a^2 + b^2} & \dfrac{a}{a^2 + b^2} \\ \dfrac{-b}{a^2 + b^2} & \dfrac{a}{a^2 + b^2} \end{pmatrix}$$

Equation $P163$, apart from providing us the general formula for the determinant of any nonsingular 2x2 matrix, is informative. It possesses a property that can be generalized to higher- order matrices to obtain their determinants. You will notice that

$$A^{-1} = \frac{cofactor\ matrix\ transposed}{|A|} = \frac{adjA}{|A|}.$$

That is, the matrix $\begin{pmatrix} d & -c \\ -b & a \end{pmatrix}$ is the transposition of the cofactor matrix which is also the adjoint matrix previously defined: $A(adjA) = (adjA)A = |A|I$.

Recall that the cofactor of any element $a_{ij} = (-1)^{i+j}minor(a_{ij})$, where $minor(a_{ij})$ is equal to the determinant of the submatrix formed by intersecting the entry a_{ij} with a horizontal and a vertical line. All the elements not on these intersecting lines form the square submatrix whose determinant gives $minor(a_{ij})$.

The cofactor matrix is defined as the matrix formed by placing the cofactor of a_{ij} at the position of a_{ij}. We then transpose the cofactor matrix to get the adjoint matrix; dividing the adjoint of A by $|A|$ finally gives A^{-1}.

Example 1.17

Find the inverse of $A = \begin{pmatrix} a & c \\ b & d \end{pmatrix}$, using the adjoint method.

Solution

The cofactor of the entry a_{ij} is

$$cofactor(a_{ij}) = (-1)^{i+j} \, minor(a_{ij})$$

$$minor(a_{11}) = \left| \begin{pmatrix} a & c \\ b & d \end{pmatrix} \right|$$

$$minor(a_{11}) = |(d)| = d$$

Recall that the determinant of a singleton matrix $S = (l)$ is $|S| = |(l)| = l$.

$$cofactor(a) = (-1)^{1+1} |(d)| = d \text{ ; similarly,}$$
$$cofactor(b) = (-1)^{2+1} |(c)| = -c$$
$$cofactor(c) = (-1)^{1+2} |(b)| = -b$$
$$cofactor(d) = (-1)^{2+2} |(a)| = a$$

$$\text{Matrix of cofactors} = \begin{pmatrix} cofactor(a) & cofactor(c) \\ cofactor(b) & cofactor(d) \end{pmatrix} = A_c$$

$$\text{Cofactor matrix } (A_c) = \begin{pmatrix} d & -b \\ -c & a \end{pmatrix}$$

$$Adj(A) = A_c^T$$

$$Adj(A) = \begin{pmatrix} d & -b \\ -c & a \end{pmatrix}^T = \begin{pmatrix} d & -c \\ -b & a \end{pmatrix}$$

Finally,

$$A^{-1} = \frac{Adj(A)}{|A|}$$

$$A^{-1} = \frac{\begin{pmatrix} d & -c \\ -b & a \end{pmatrix}}{ad-bc}, \text{ confirming the formula.}$$

Therefore, for any nxn matrix A,

$$A^{-1} = \frac{A_c^T}{|A|} \; ; |A| \neq 0 \qquad\qquad ----- P164$$

To see that $A^{-1} = \frac{adj(A)}{|A|}$, analytically, from the definition,

$$A[adj(A)] = adj(A)A = |A|I \text{ , we have}$$
$$A[adj(A)] = |A|I$$

Left multiplying both sides by A^{-1}, we have

$$A^{-1}A[adj(A)] = A^{-1}|A|I$$
$$\Rightarrow \quad Iadj(A) = |A| \, A^{-1}I$$
$$adjA = |A|A^{-1}$$
$$\Rightarrow \quad A^{-1} = \frac{adj(A)}{|A|}.$$

Example 1.18

Using the inverse method, solve the system

$3x_1 - x_2 + 2x_3 = -1$

$-x_1 + 5x_2 + 10x_3 = 13$

$2x_1 - 3x_2 - 8x_3 = -3$

Solution

The coefficient matrix A is

$$A = \begin{pmatrix} 3 & -1 & 2 \\ -1 & 5 & 10 \\ 2 & -3 & -8 \end{pmatrix}$$

$$cofactor(a_{11}) = (-1)^{1+1} \begin{vmatrix} 5 & 10 \\ -3 & -8 \end{vmatrix} = -10$$

$$cofactor(a_{12}) = (-1)^{1+2}\begin{vmatrix} -1 & 10 \\ 2 & -8 \end{vmatrix} = 12$$

$$cofactor(a_{13}) = (-1)^{1+3}\begin{vmatrix} -1 & 2 \\ 2 & -3 \end{vmatrix} = -7$$

$$cofactor(a_{21}) = (-1)^{2+1}\begin{vmatrix} -1 & 2 \\ -3 & -8 \end{vmatrix} = -14$$

$$cofactor(a_{22}) = (-1)^{2+2}\begin{vmatrix} 3 & 2 \\ 2 & -8 \end{vmatrix} = -20$$

$$cofactor(a_{23}) = (-1)^{2+3}\begin{vmatrix} 3 & -1 \\ 2 & -3 \end{vmatrix} = 7$$

$$cofactor(a_{31}) = (-1)^{3+1}\begin{vmatrix} -1 & 2 \\ 5 & 10 \end{vmatrix} = -20$$

$$cofactor(a_{32}) = (-1)^{3+2}\begin{vmatrix} 3 & 2 \\ -1 & 10 \end{vmatrix} = -32$$

$$cofactor(a_{33}) = (-1)^{3+3}\begin{vmatrix} 3 & -1 \\ -1 & 5 \end{vmatrix} = 14$$

$$A_c = \begin{pmatrix} -10 & 12 & -7 \\ -14 & -20 & 7 \\ -20 & -32 & 14 \end{pmatrix}$$

$$A_c^T = \begin{pmatrix} -10 & -14 & -20 \\ -12 & -20 & -32 \\ -7 & 7 & 14 \end{pmatrix}$$

$$|A| = \begin{vmatrix} 3 & -1 & 2 \\ -1 & 5 & 10 \\ 2 & -3 & -8 \end{vmatrix},$$ expanding along the first column, we have

$$|A| = (-1)^{1+1}.3(-40 + 30) + (-1)^{2+1}.(-1)(8 + 6) + (-1)^{3+1}.2(-10 - 10)$$

$$|A| = -30 + 14 - 40 = -56$$

$$A^{-1} = \frac{A_c^T}{|A|}$$

$$A^{-1} = \frac{-1}{56}\begin{pmatrix} -10 & -14 & -20 \\ 12 & -20 & -32 \\ -7 & 7 & 14 \end{pmatrix}$$

$$X = A^{-1}B$$

$$\begin{pmatrix} x_1 \\ x_2 \\ x_3 \end{pmatrix} = \frac{-1}{56}\begin{pmatrix} -10 & -14 & -20 \\ 12 & -20 & -32 \\ -7 & 7 & 14 \end{pmatrix}\begin{pmatrix} -1 \\ 13 \\ -3 \end{pmatrix}$$

$$\begin{pmatrix} x_1 \\ x_2 \\ x_3 \end{pmatrix} = \frac{-1}{56}\begin{pmatrix} 10 - 182 + 60 \\ -12 - 260 + 96 \\ 7 + 91 - 42 \end{pmatrix} = \frac{-1}{56}\begin{pmatrix} -112 \\ -344 \\ 56 \end{pmatrix}$$

$$\Rightarrow \begin{pmatrix} x_1 \\ x_2 \\ x_3 \end{pmatrix} = \begin{pmatrix} 2 \\ 5 \\ -1 \end{pmatrix}$$

$$\Rightarrow x_1 = 2 \,, x_2 = 5 \,, x_3 = -1$$

Row-transformation method can also be used. In fact, the inverse formula was derived from the row-transformation method as we saw it.

The Method of Determinants: Crammer's Rule

This method is called the method of determinant because it strictly involves evaluating determinants; no inverse. Consider the system

$$AX = B$$

If A is an nxn matrix, and $A \neq 0$, then we need to evaluate $n + 1$ determinants; One of them is $|A|$ itself. If $x_1, x_2, \ldots x_n$ are the unknowns, then we need to evaluate, in addition to $|A|$, n other determinants

$$|A(x_1)|, |A(x_2)|, \ldots\ldots |A(x_n)|$$

The matrix $A(x_i)$ is formed by replacing the column vector of x_i in A with the constant vector B, after which each x_i is evaluated using the formula

$$x_i = \frac{|A(x_i)|}{|A|}$$

Note that (x_i) does not mean a singleton matrix here; it connotes something different.

Example 1.19:

Using crammer's rule, solve the linear system

$$2x_1 + 3x_2 = 8$$
$$3x_1 + 5x_2 = 31$$

Solution

$$A = \begin{vmatrix} 2 & 3 \\ 3 & 5 \end{vmatrix} = 19$$

$$|A(x_1)| = \begin{vmatrix} 8 & -3 \\ 31 & 5 \end{vmatrix} = 133$$

$$|A(x_2)| = \begin{vmatrix} 2 & 8 \\ 3 & 31 \end{vmatrix} = 38$$

$$x_1 = \frac{|A(x_1)|}{|A|} = \frac{133}{19} = 7$$

$$x_2 = \frac{|A(x_2)|}{|A|} = \frac{38}{19} = 2$$

Example 1.20:

Solve the following systems, using crammer's rule in each case.

a) $x_1 + x_2 - x_3 = 1$
   $x_1 - 2x_2 + x_3 = 2$
   $2x_1 + 3x_2 - 3x_3 = 1$

b) $x_1 - x_2 + 2x_3 - x_4 = 6$
   $2x_1 + 3x_2 + 4x_3 - 2x_4 = 7$
   $-x_1 + x_2 + 2x_3 + x_4 = -4$
   $6x_1 + 2x_2 + x_4 = 1$

c) $3x_1 - x_2 + x_4 = 4$
   $-x_1 + 3x_2 - x_3 = -2$
   $x_1 + x_2 + 2x_3 + x_4 = 8$
   $5x_1 + 3x_2 - 2x_3 + 2x_5 = 5$

Solution

a) $AX = B$

$$\begin{pmatrix} 1 & 1 & -1 \\ 1 & -2 & 1 \\ 2 & 3 & -3 \end{pmatrix} \begin{pmatrix} x_1 \\ x_2 \\ x_3 \end{pmatrix} = \begin{pmatrix} 1 \\ 2 \\ 1 \end{pmatrix}$$

We can expand directly to find determinant, but to illustrate some of the properties of determinant, let us simplify before expanding.

$$|A| = \begin{vmatrix} 1 & 1 & -1 \\ 1 & -2 & 1 \\ 2 & 3 & -3 \end{vmatrix}$$

$$R_1 + R_2$$

$$|A| = \begin{vmatrix} 2 & -1 & 0 \\ 1 & -2 & 1 \\ 2 & 3 & -3 \end{vmatrix}$$

$$(R_3 + 3R_2)$$

$$|A| = \begin{vmatrix} 2 & -1 & 0 \\ 1 & -2 & 1 \\ 5 & -3 & 0 \end{vmatrix}$$

Expanding along the third column,

$$|A| = (-1)^{2+3} (1)(-6 + 5)$$

$$|A| = 1$$

$$|A(x_1)| = \begin{vmatrix} 1 & 1 & -1 \\ 2 & 2 & 1 \\ 1 & 3 & -3 \end{vmatrix}$$

$$(R_2 - 2R_1), (R_3 - R_1)$$

$$|A(x_1)| = \begin{vmatrix} 1 & 1 & -1 \\ 0 & -4 & 3 \\ 0 & 2 & -2 \end{vmatrix}$$

$$(R_2 \sim R_3), (R_3 + 2R_2)$$

$$|A(x_1)| = -\begin{vmatrix} 1 & 1 & -1 \\ 0 & 2 & -2 \\ 0 & 0 & -1 \end{vmatrix} = -(1.2.-1)$$

Recall that the determinant of a triangular matrix is equal to the product of its main diagonal entries.

$$|A(x_1)| = 2$$

$$|A(x_2)| = \begin{vmatrix} 1 & 1 & -1 \\ 1 & 2 & 1 \\ 2 & 1 & -3 \end{vmatrix}$$

Expanding along the first row

$$|A(x_2)| = (1)(-6 - 1) - 1(-3 - 2) + -1(1 - 4)$$

$$|A(x_2)| = -7 + 5 - 3$$

$$|A(x_2)| = 1$$

$$|A(x_3)| = \begin{vmatrix} 1 & 1 & 1 \\ -1 & 2 & 2 \\ 2 & 3 & 1 \end{vmatrix}$$

Expanding along the first row,

$$|A(x_3)| = (1)(-2-6) - 1(1-4) + 1(3+4)$$
$$|A(x_3)| = -8 + 3 + 7 = 2$$

$$x_1 = \frac{|A(x_1)|}{|A|} = \frac{2}{1} = 2$$

$$x_2 = \frac{|A(x_2)|}{|A|} = \frac{1}{1} = 1$$

$$x_3 = \frac{|A(x_3)|}{|A|} = \frac{2}{1} = 2$$

$$\begin{pmatrix} x_1 \\ x_2 \\ x_3 \end{pmatrix} = \begin{pmatrix} 2 \\ 1 \\ 2 \end{pmatrix}$$

b)
$$\begin{pmatrix} 1 & -1 & 2 & -1 \\ 2 & 3 & 4 & -2 \\ -1 & 1 & 2 & 1 \\ 6 & 2 & 0 & 1 \end{pmatrix} \begin{pmatrix} x_1 \\ x_2 \\ x_3 \\ x_4 \end{pmatrix} = \begin{pmatrix} 6 \\ 7 \\ -4 \\ 1 \end{pmatrix}$$

$$|A| = \begin{vmatrix} 1 & -1 & 2 & -1 \\ 2 & 3 & 4 & -2 \\ -1 & 1 & 2 & 1 \\ 6 & 2 & 0 & 1 \end{vmatrix}$$

Expanding directly implies that we will have to evaluate 3x3 determinants four times. We therefore take advantage of determinant properties together with row operations in order to simplify the determinants.

$$(R_4 + R_1), (R_3 + R_1), (R_2 - 2R_1)$$

$$|A| = \begin{vmatrix} 1 & -1 & 2 & -1 \\ 0 & 5 & 0 & 0 \\ 0 & 0 & 4 & 0 \\ 7 & 1 & 2 & 0 \end{vmatrix}$$

Expanding along the 4th column,

$$|A| = (-1)^{1+4}(-1)\begin{vmatrix} 0 & 5 & 0 \\ 0 & 0 & 4 \\ 7 & 1 & 2 \end{vmatrix} = \begin{vmatrix} 0 & 5 & 0 \\ 0 & 0 & 4 \\ 7 & 1 & 2 \end{vmatrix}$$

Expanding further along the 2nd row,

$$|A| = (-1)^{2+2}(4)\begin{vmatrix} 0 & 5 \\ 7 & 1 \end{vmatrix}$$

$$= 4(0 - 35)$$

$$= -140$$

$$|A(x_1)| = \begin{vmatrix} 6 & -1 & 2 & -1 \\ 7 & 3 & 4 & -2 \\ -4 & 1 & 2 & 1 \\ 1 & 2 & 0 & 1 \end{vmatrix}$$

$$(R_4 + R_1), (R_3 + R_1), (R_2 - 2R_1)$$

$$|A(x_1)| = \begin{vmatrix} 6 & -1 & 2 & -1 \\ -5 & 5 & 0 & 0 \\ 2 & 0 & 4 & 0 \\ 7 & 1 & 2 & 0 \end{vmatrix}$$

Expanding along the fourth column,

$$|A(x_1)| = (-1)^{1+4}(-1)\begin{vmatrix} -5 & 5 & 0 \\ 2 & 0 & 4 \\ 7 & 1 & 2 \end{vmatrix}$$

$$|A(x_1)| = \begin{vmatrix} -5 & 5 & 0 \\ 2 & 0 & 4 \\ 7 & 1 & 2 \end{vmatrix}$$

$$(R_2 - 2R_3)$$

$$|A(x_1)| = \begin{vmatrix} -5 & 5 & 0 \\ -12 & -2 & 0 \\ 7 & 1 & 2 \end{vmatrix}$$

Expanding along the third column,

$$|A(x_1)| = (-1)^6(2) \begin{vmatrix} -5 & 5 \\ -12 & -2 \end{vmatrix} = 2(10+60)$$

$$|A(x_1)| = 14$$

$$|A(x_2)| = \begin{vmatrix} 1 & 6 & 2 & -1 \\ 2 & 7 & 4 & -2 \\ -1 & -4 & 2 & 1 \\ 6 & 1 & 0 & 1 \end{vmatrix}$$

$$(R_4 + R_1), (R_3 + R_1), (R_2 - 2R_1)$$

$$|A(x_2)| = \begin{vmatrix} 1 & 6 & 2 & -1 \\ 0 & -5 & 0 & 0 \\ 0 & 2 & 4 & 0 \\ 7 & 7 & 2 & 0 \end{vmatrix}$$

Expanding along the 4th column

$$|A(x_2)| = (-1)^5(-1) \begin{vmatrix} 0 & -5 & 0 \\ 0 & 2 & 4 \\ 7 & 7 & 2 \end{vmatrix}$$

$$|A(x_2)| = \begin{vmatrix} 0 & -5 & 0 \\ 0 & 2 & 4 \\ 7 & 7 & 2 \end{vmatrix}$$

Expanding along the 1st column

$$|A(x_2)| = (-1)^{3+1}(7) \begin{vmatrix} -5 & 0 \\ 2 & 4 \end{vmatrix} = 7(-20)$$

$$|A(x_2)| = -140$$

$$|A(x_3)| = \begin{vmatrix} 1 & -1 & 6 & -1 \\ 2 & 3 & 7 & -2 \\ -1 & 1 & -4 & 1 \\ 6 & 2 & 1 & 1 \end{vmatrix}$$

$$(R_4 + R_1), (R_3 + R_1), (R_2 - 2R_1)$$

$$|A(x_3)| = \begin{vmatrix} 1 & -1 & 6 & -1 \\ 0 & 5 & -5 & 0 \\ 0 & 0 & 2 & 0 \\ 7 & 1 & 7 & 0 \end{vmatrix}$$

Expanding along the fourth column,

$$|A(x_3)| = (-1)^5(-1) \begin{vmatrix} 0 & 5 & -5 \\ 0 & 0 & 2 \\ 7 & 1 & 7 \end{vmatrix}$$

Expanding along the second row,

$$|A(x_3)| = (-1)^{2+3}(2) \begin{vmatrix} 0 & 5 \\ 7 & 1 \end{vmatrix}$$

$$= -2(-35)$$

$$|A(x_3)| = 70$$

Finally,

$$|A(x_4)| = \begin{vmatrix} 1 & -1 & 2 & 6 \\ 2 & 3 & 4 & 7 \\ -1 & 1 & 2 & -4 \\ 6 & 2 & 0 & 1 \end{vmatrix}$$

$$(R_3 - R_1), (R_2 - 2R_1)$$

$$|A(x_4)| = \begin{vmatrix} 1 & -1 & 2 & 6 \\ 0 & 5 & 0 & -5 \\ -2 & 2 & 0 & -10 \\ 6 & 2 & 0 & 1 \end{vmatrix}$$

Expanding along the third column,

$$|A(x_4)| = (-1)^{1+3}(2) \begin{vmatrix} 0 & 5 & -5 \\ -2 & 2 & -10 \\ 6 & 2 & 1 \end{vmatrix}$$

$$(R_3 + 3R_2)$$

$$|A(x_4)| = 2 \begin{vmatrix} 0 & 5 & -5 \\ -2 & 2 & -10 \\ 0 & 8 & -29 \end{vmatrix}$$

Expanding along the first column,

$$|A(x_4)| = 2.(-1)^{2+1}(-2) \begin{vmatrix} 5 & -5 \\ 8 & -29 \end{vmatrix}$$

$$= 4(-145 + 40)$$

$$|A(x_4)| = -420$$

$$x_1 = \frac{|A(x_1)|}{|A|} = \frac{140}{-140} = -1$$

$$x_2 = \frac{|A(x_2)|}{|A|} = \frac{-140}{-140} = 1$$

$$x_3 = \frac{|A(x_3)|}{|A|} = \frac{-70}{-140} = \frac{-1}{2}$$

$$x_4 = \frac{|A(x_4)|}{|A|} = \frac{-420}{-140} = 3$$

$$\begin{pmatrix} x_1 \\ x_2 \\ x_3 \\ x_4 \end{pmatrix} = \begin{pmatrix} -1 \\ 1 \\ \frac{-1}{2} \\ 3 \end{pmatrix}$$

c)
$$AX = B$$

$$\begin{pmatrix} 3 & -1 & 0 & 1 \\ -1 & 3 & -1 & 0 \\ 1 & 1 & 2 & 1 \\ 5 & 3 & -2 & 2 \end{pmatrix} \begin{pmatrix} x_1 \\ x_2 \\ x_3 \\ x_4 \end{pmatrix} = \begin{pmatrix} 4 \\ -2 \\ 8 \\ 5 \end{pmatrix}$$

$$|A| = \begin{vmatrix} 3 & -1 & 0 & 1 \\ -1 & 3 & -1 & 0 \\ 1 & 1 & 2 & 1 \\ 5 & 3 & -2 & 2 \end{vmatrix}$$

Since $|A| = |A^T|$, transposing we have

$$|A| = \begin{vmatrix} 3 & -1 & 1 & 5 \\ -1 & 3 & 1 & 3 \\ 0 & -1 & 2 & -2 \\ 1 & 0 & 1 & 2 \end{vmatrix}$$

$$(R_3 - 2R_1), (R_2 - R_1), (R_4 - R_1)$$

$$|A| = \begin{vmatrix} 3 & -1 & 1 & 5 \\ -4 & 4 & 0 & -2 \\ -6 & 1 & 0 & -12 \\ -2 & 1 & 0 & -3 \end{vmatrix}$$

$$|A| = (-1)^{1+3}(1) \begin{vmatrix} -4 & 4 & -2 \\ -6 & 1 & -12 \\ -2 & 1 & -3 \end{vmatrix}$$

$$|A| = \begin{vmatrix} -4 & 4 & -2 \\ -6 & 1 & -12 \\ -2 & 1 & -3 \end{vmatrix}$$

$$\tfrac{1}{4}(R_1)$$

$$|A| = 4 \begin{vmatrix} -1 & 1 & \frac{-1}{2} \\ -6 & 1 & -12 \\ -2 & 1 & -3 \end{vmatrix}$$

$$(R_2 - R_1)(R_3 - R_1)$$

$$|A| = 4 \begin{vmatrix} -1 & 1 & \dfrac{-1}{2} \\ -5 & 0 & \dfrac{-23}{2} \\ 4 & 0 & 9 \end{vmatrix}$$

$$|A| = 4(-1)^{1+2}(1) \begin{vmatrix} -5 & \dfrac{-23}{2} \\ 4 & 9 \end{vmatrix}$$

$$|A| = -4\left(-45 + 4\left(\frac{23}{2}\right)\right)$$

$$|A| = -4$$

$$|A(x_1)| = \begin{vmatrix} 4 & -1 & 0 & 1 \\ -2 & 3 & -1 & 0 \\ 8 & 1 & 2 & 1 \\ 5 & 3 & -2 & 2 \end{vmatrix}$$

$$(R_3 - R_1), (R_4 - 2R_1)$$

$$|A(x_1)| = \begin{vmatrix} 4 & -1 & 0 & 1 \\ -2 & 3 & -1 & 0 \\ 4 & 2 & 2 & 0 \\ -3 & 5 & -2 & 0 \end{vmatrix}$$

$$= (-1)^5(1) \begin{vmatrix} -2 & 3 & -1 \\ 4 & 2 & 2 \\ -3 & 5 & -2 \end{vmatrix}$$

$$(R_3 + R_2), (R_2 + 2R_1)$$

$$|A(x_1)| = - \begin{vmatrix} -2 & 3 & -1 \\ 0 & 8 & 0 \\ 1 & 7 & 0 \end{vmatrix}$$

$$|A(x_1)| = -(-1)^{2+2}(8) \begin{vmatrix} -2 & -1 \\ 1 & 0 \end{vmatrix}$$

$$|A(x_1)| = -8$$

$$|A(x_2)| = \begin{vmatrix} 3 & 4 & 0 & 1 \\ -1 & -2 & -1 & 0 \\ 1 & 8 & 2 & 1 \\ 5 & 5 & -2 & 2 \end{vmatrix}$$

$$(R_3 - R_1), (R_4 + 2R_1)$$

$$|A(x_2)| = \begin{vmatrix} 3 & 4 & 0 & 1 \\ -1 & -2 & -1 & 0 \\ 2 & 4 & 2 & 0 \\ -1 & -3 & -2 & 0 \end{vmatrix}$$

$$= (-1)^{1+4}\,(1) \begin{vmatrix} -1 & -2 & -1 \\ -2 & 4 & 2 \\ -1 & -3 & -2 \end{vmatrix}$$

$$(R_3 + R_2),\ (R_2 + 2R_1)$$

$$|A(x_2)| = -\begin{vmatrix} -1 & -2 & -1 \\ -4 & 0 & 0 \\ -3 & 1 & 0 \end{vmatrix} = -(-1)^{1+3}\,(-1)\begin{vmatrix} -4 & 0 \\ -3 & 1 \end{vmatrix}$$

$$|A(x_2)| = -4$$

$$|A(x_3)| = \begin{vmatrix} 3 & -1 & 4 & 1 \\ -1 & 3 & -2 & 0 \\ 1 & 1 & 8 & 1 \\ 5 & 3 & 5 & 2 \end{vmatrix}$$

$$(R_3 - R_1),\ (R_4 - 2R_1)$$

$$|A(x_3)| = \begin{vmatrix} 3 & -1 & 4 & -1 \\ -1 & 3 & -2 & 0 \\ -2 & 2 & 4 & 0 \\ -1 & 5 & -3 & 0 \end{vmatrix}$$

$$= (-1)1+4\,(-1)\begin{vmatrix} -1 & 3 & -2 \\ -2 & 2 & 4 \\ -1 & 5 & -3 \end{vmatrix}$$

$$(R_3 - R_1),\ (R_2 - 2R_1)$$

$$|A(x_3)| = \begin{vmatrix} -1 & 3 & -2 \\ 0 & -4 & 8 \\ 0 & 2 & -1 \end{vmatrix}$$

$$|A(x_3)| = 12$$

$$|A(x_4)| = \begin{vmatrix} 3 & -1 & 0 & 4 \\ -1 & 3 & -1 & -2 \\ 1 & 1 & 2 & 8 \\ 5 & 3 & -2 & 5 \end{vmatrix}$$

$$(R_3 + 2R_2), (R_4 - 2R_2)$$

$$|A(x_4)| = \begin{vmatrix} 3 & -1 & 0 & 4 \\ -1 & 3 & -1 & -2 \\ -1 & 7 & 0 & 4 \\ 7 & -3 & 0 & 9 \end{vmatrix}$$

$$= (-1)^{2+3}(-1) \begin{vmatrix} 3 & -1 & 4 \\ -1 & 7 & 4 \\ 7 & -3 & 9 \end{vmatrix}$$

Expanding along the first row,

$$|A(x_4)| = 3\,(63 + 12) + 1\,(-9 - 28) + 4(3 - 49)$$

$$|A(x_4)| = 4$$

$$x_1 = \frac{A(x_1)}{A} = \frac{8}{4} = 2\,, x_2 = \frac{|A(x_2)|}{|A|} = \frac{4}{4} = 1\,, x_3 = \frac{|A(x_3)|}{|A|} = \frac{12}{4} = 3,$$

$$x_4 = \frac{|A(x_4)|}{A} = \frac{4}{4} = 1$$

$$\therefore \begin{pmatrix} x_1 \\ x_2 \\ x_3 \\ x_4 \end{pmatrix} = \begin{pmatrix} 2 \\ 1 \\ -3 \\ -1 \end{pmatrix}$$

Example 1.21:

$$\sqrt{x} - \sqrt{y} + \sqrt{z} = 3$$

$$2\sqrt{x} + \sqrt{y} - \sqrt{z} = 3$$

$$\sqrt{x} - \sqrt{y} + 2\sqrt{z} = 7$$

Solution

This system of equation is not linear. It can be made linear by making the following simple substitutions.

Let

$$x_1 = \sqrt{x}, x_2 = \sqrt{y}, x_3 = \sqrt{z}$$

The system transforms to the linear system

a) $\quad x_1 - x_2 + x_3 = 3$

$\quad\quad 2x_1 + x_2 - x_3 = 3$

$\quad\quad x_1 - x_2 + 2x_3 = 7$

Let us use the row transformation method.

$$\left(\begin{array}{ccc|ccc} 1 & -1 & 1 & 1 & 0 & 0 \\ 2 & 1 & -1 & 0 & 1 & 0 \\ 1 & -1 & 2 & 0 & 0 & 1 \end{array}\right) (R_2 - 2R_1), (R_3 - R_1) \underset{\sim}{\quad} \left(\begin{array}{ccc|ccc} 1 & -1 & 1 & 1 & 0 & 0 \\ 0 & 3 & -3 & -2 & 1 & 0 \\ 0 & 0 & 1 & -1 & 0 & 1 \end{array}\right)$$

$$(\tfrac{1}{3}R_2), (R_1 + R_2) \underset{\sim}{\quad} \left(\begin{array}{ccc|ccc} 1 & 0 & 0 & \frac{1}{3} & \frac{1}{3} & 0 \\ 0 & 1 & -1 & \frac{-2}{3} & \frac{1}{3} & 0 \\ 0 & 0 & 1 & \frac{3}{-1} & 0 & 1 \end{array}\right) (R_2 + R_3) \underset{\sim}{\quad} \left(\begin{array}{ccc|ccc} 1 & 0 & 0 & \frac{1}{3} & \frac{1}{3} & 0 \\ 0 & 1 & 0 & \frac{-5}{3} & \frac{1}{3} & 1 \\ 0 & 0 & 1 & -1 & 0 & 1 \end{array}\right)$$

The inverse of the coefficient matrix is

$$A^{-1} = \frac{1}{3}\begin{pmatrix} 1 & 1 & 0 \\ -5 & 1 & 3 \\ -3 & 0 & 3 \end{pmatrix}$$

$$\begin{pmatrix} x_1 \\ x_2 \\ x_3 \end{pmatrix} = A^{-1}B$$

where,

$$B = \begin{pmatrix} 3 \\ 3 \\ 7 \end{pmatrix}$$

$$\Rightarrow \qquad \begin{pmatrix} x_1 \\ x_2 \\ x_3 \end{pmatrix} = \frac{1}{3} \begin{pmatrix} 1 & 1 & 0 \\ -5 & 1 & 3 \\ -3 & 0 & 3 \end{pmatrix} \begin{pmatrix} 3 \\ 3 \\ 7 \end{pmatrix}$$

$$\begin{pmatrix} x_1 \\ x_2 \\ x_3 \end{pmatrix} = \frac{1}{3} \begin{pmatrix} 3 & + & 3 & + & 0 \\ -15 & + & 3 & + & 21 \\ -9 & + & 0 & + & 21 \end{pmatrix}$$

$$\begin{pmatrix} x_1 \\ x_2 \\ x_3 \end{pmatrix} = \begin{pmatrix} 2 \\ 3 \\ 4 \end{pmatrix}$$

$$\Rightarrow \qquad x_1 = 2 \,, x_2 = 3 \,, x_3 = 4$$

$$\sqrt{x} = 2, \ \sqrt{y} = 3, \ \sqrt{z} = 4$$

$$\Rightarrow \qquad x = 4, y = 9, z = 16$$

Example 1.22:

$x^3 - 3y^3 - z^3 = -4$

$2x^3 - y^3 - z^3 = 9$

$5x^3 - 3y^3 - z^3 = 0$

Solution

Again, this equation is nonlinear, but can be made linear by the substitutions

$$x_1 = x^3 \,, \ x_2 = y^3 \,, \ x_3 = z^3$$

a) $\quad x_1 - 3x_2 - x_3 = -4$

$\qquad 2x_1 + x_2 + x_3 = 9$

$\qquad 5x_1 - 3x_2 - x_3 = 0$

Using Gaussian elimination, we have

$$\begin{pmatrix} 1 & -3 & -1 & | & -4 \\ 2 & 1 & 1 & | & 9 \\ 5 & -3 & -1 & | & 0 \end{pmatrix} (R_3 - 5R_1), (R_2 - 2R_1) \sim \begin{pmatrix} 1 & -3 & -1 & | & -4 \\ 0 & 7 & 3 & | & 17 \\ 0 & 12 & 4 & | & 20 \end{pmatrix}$$

$$\sim (R_2 \sim R_3), (\tfrac{1}{12}R_2)\begin{pmatrix} 1 & -3 & -1 & -4 \\ 0 & 1 & \tfrac{1}{3} & \tfrac{5}{3} \\ 0 & 7 & 3 & 17 \end{pmatrix} \sim (R_3 - 7R_2)\begin{pmatrix} 1 & -3 & -1 & -4 \\ 0 & 1 & \tfrac{1}{3} & \tfrac{5}{3} \\ 0 & 0 & \tfrac{2}{3} & \tfrac{16}{3} \end{pmatrix}$$

$$\Rightarrow \quad \tfrac{2}{3}x_3 = \tfrac{16}{3}, \, x_3 = 8, \, x_2 + \tfrac{1}{3}x_3 = \tfrac{5}{3}, \, x_2 = -1$$

$$x_1 - 3x_2 - x_3 = -4, \, x_1 = 1$$

$$\begin{pmatrix} x_1 \\ x_2 \\ x_3 \end{pmatrix} = \begin{pmatrix} 1 \\ -1 \\ 8 \end{pmatrix}$$

$$\Rightarrow x^3 = 1, y^3 = -1, z^3 = 8$$

$$x = 1, y = -1, z = 2$$

Chapter Two

Vector Spaces

Central to much of what will be covered in the future is the idea of vector spaces and related concepts; thus, it is necessary that we dedicate the next few pages of this text to a brief treatment of these concepts.

Let V be a vector space over the field R whose elements are real numbers. We say that V is a real vector space if all of the following axioms hold:

Axioms for a real vector space

i)   Closure law: If v_1 and v_2 belong to V, then $v_1 + v_2$ is also in V and is unique.

ii)   Commutativity: If v_1 and v_2 belong to V, then $v_1 + v_2 = v_2 + v_1$.

iii)   Associativity: for v_1, v_2, v_3 in V, $(v_1 + v_2) + v_3 = v_1 + (v_2 + v_3)$

iv)   Additive Identity: There is always a vector 0 in V called the zero vector such that for every v in V, $v + 0 = 0 + v = v$

v)   Additive Inverse: for every v in V, there is a vector $-v$, such that $v + (-v) = -v + v = 0$. The vector $-v$ is called the negative of the vector v.

Axioms for scalar multiplication

i)   If p is any scalar in R and v belongs to V, then the vector $p.v = pv$ is also in V and is unique.

ii)   For every p in R, and v_1, v_2 in V,
$p(v_1 + v_2) = pv_1 + pv_2$

iii)   For scalars p_1, p_2, in R, and vector v in V,
$(p_1 + p_2)v = p_1v + p_1v$, and $(p_1 p_2)v = p_1(p_2)v = p_2(p_1 v)$

iv) $1. v = v$ for every v in V.

[Note: We say that the vector space V is a complex-vector space or a vector space over C if the set of real numbers R is replaced with the set of complex numbers C in the axioms for vector space.]

Certain examples of a real vector space

Example

Show that the set of all points (ordered pairs) in the real coordinate plane is a real vector space.

Solution

We simply want to show that the set of all ordered pairs (x_1, x_2), where x_1 and x_2 are real numbers in the real coordinate plane, obeys all the axioms of a real vector space; we will only show some of them, leaving out the highly obvious ones as exercises for the readers. Let V be the set of all ordered pairs in the real coordinate plane.

Closure law: for any two ordered pairs (x_1, x_2) and (x_1', x_2') in the plane, the point $(x_1 + x_1', x_2 + x_2')$ is also an element of the real plane.

The commutative and the associative laws follow directly as a result of the commutativity and associativity of real numbers.

Additive Identity: in the real plane, there is the vector $(0,0)$ called the origin. The point $(0,0)$ is the additive identity because for any (x_1, x_2) in V,

$$(x_1, x_2) + (0,0) = (x_1 + 0, x_2 + 0) = (x_1, x_2)$$
$$(0,0) + (x_1, x_2) = (0 + x_1, 0 + x_2) = (x_1, x_2)$$

Additive Inverse: for every (x_1, x_2) in V, there is obviously the point $(-x_1, -x_2)$ in V such that

$$(x_1, x_2) + (-x_1, -x_2) = (x_1 + -x_1, x_2 + -x_2) = (x_1 - x_1, x_2 - x_2) = (0,0)$$

Again,

$$(-x_1, -x_2) + (x_1, x_2) = (-x_1 + x_1, -x_2 + x_2) = (0,0).$$

For every scalar p, and every vector (x_1, x_2) in V, there is the vector

$p(x_1, x_2) = (px_1, px_2)$; also, $p_1 p_2 (x_1, x_2) = p_1(p_2 x_1, p_2 x_2)$ for every scalar p_1 and p_2.

It is easy to verify other axioms. We leave it as an exercise for the reader.

Note that the column vectors of any 2xn , real matrix are members of (or belong to) R^2; $n \geq 1$, nϵ N.

Example 2.1

Show that the set of all ordered triples (x_1, x_2, x_3) in 3-space, R^3, is a vector space.

Solution

Let V be the set of all ordered triples, (x_1, x_2, x_3) in R^3, then

Closure: for **every** (x_1, x_2, x_3) and (x_1', x_2', x_3') in V,

$$(x_1 + x_1', x_2 + x_2', x_3 + x_3') \text{ is also in V.}$$

If p is any scalar, and (x_1, x_2, x_3) is in V, then the vector

$$p(x_1, x_2, x_3) = (px_1, px_2, px_3) \text{ is in V.}$$

The zero vector (0, 0, 0) is in V, and for every (x_1, x_2, x_3) in V, there is $(-x_1, -x_2, -x_3)$ in V. with these properties shown, it is easy to establish all the other axioms for V.

Again, note that the column vectors of any 3xn real matrix are from R^3; $n \geq 1$, nϵ N.

Example 2.2

Show that the set P_n of all polynomials of degree n or less with real coefficients together with the zero polynomial is a real vector space.

Solution

Any polynomial $f(x)$ of degree n or less can always be written in the form

$$f(x) = \sum_{i=0}^{n} a_i x^i; \quad a_i \text{ belongs to R}$$

If $g(x) = \sum_{i=0}^{n} b_i x^i$, b_i *belongs to* R, then

$$f(x) + g(x) = \sum_{i=0}^{n} a_i x^i + \sum_{i=0}^{n} b_i x^i$$

$$= \sum_{i=0}^{n} (a_i + b_i) x^i \text{, which is also in } P_n.$$

Thus, the closure law is established.

For any scalar q in R, and $f(x)$ in P_n, the vector (polynomial)

$$q f(x) = q \sum_{i=0}^{n} a_i x^i = \sum_{i=0}^{n} q a_i x^i \text{ is also in } p_n.$$

Also, it is easy to see that for every $f(x)$ in p_n, the vector (polynomial)

$$-f(x) = -\sum_{i=0}^{n} a_i x^i = \sum_{i=0}^{n} -a_i x^i = \text{ is in } p_n.$$

Since the zero polynomial $f(x) = 0$ is in p_n, therefore every other axiom follows from these established ones. Thus, p_n is a vector space.

Example 2.3

Show that the set M_n of all $n \times n$ matrices with real entries, together with the $n \times n$ null matrix, forms a vector space.

Solution

For vectors $v_1 = \begin{pmatrix} x_{11} & x_{12} \dots \dots & x_{1n} \\ x_{21} & x_{22} \dots \dots & x_{2n} \\ \vdots & \vdots & \vdots \\ \vdots & \vdots & \vdots \\ x_{n1} & x_{n2} & x_{nn} \end{pmatrix}$, $v_2 = \begin{pmatrix} x'_{11} & x'_{12} \dots \dots & x'_{1n} \\ x'_{21} & x'_{22} \dots \dots & x'_{2n} \\ \vdots & \vdots & \vdots \\ \vdots & \vdots & \vdots \\ x'_{n1} & x'_{n2} \dots \dots \dots & x'_{nn} \end{pmatrix}$ in M_n,

The vector $v_1 + v_2 = \begin{pmatrix} x_{11} + x'_{11} & x_{12} + x'_{12} \dots \dots & x_{1n} + x'_{1n} \\ x_{21} + x'_{21} & x_{22} + x'_{22} \dots \dots \dots & x_{2n} + x'_{2n} \\ \vdots & \vdots & \vdots \\ \vdots & \vdots & \vdots \\ x_{n1} + x'_{n1} & x_{n1} + x'_{n1} \dots \dots \dots & x_{nn} + x'_{nn} \end{pmatrix}$ is also in M_n

For every v_1 in M_n, the vector qv_1 is in M_n, where q is any scalar; also the vector $-v_1$ is in M_n. Since the $n \times n$ null matrix is in M_n; therefore, M_n is a vector space.

Example 2.4

Show that the solution space of the linear homogeneous system $AX = 0$, denoted as S, is a vector space.

Solution

The solution to $AX = 0$ is either unique, in which case the only solution is the zero vector $v = 0$, or non unique, in which case there is an infinite number of solutions.

We first assume non-uniqueness for the system. If x_1 and x_2 are two solutions to the homogenous system

$AX = 0$; that is, if x_1 and x_2 belong to the solution space S, then $x_1 + x_2$ is also in S. This is because

$$Ax_1 = 0; \text{ since } x_1 \text{ belongs to S.}$$
$$Ax_2 = 0; \text{ since } x_2 \text{ belongs to S.}$$
$$Ax_1 + Ax_2 = 0$$
$$\Rightarrow \quad A(x_1 + x_2) = 0$$

meaning that $x_1 + x_2$ is also a solution whenever x_1 and x_2 are; that is, $x_1 + x_2$ is also in S whenever x_1 and x_2 are.

Again, for any scalar q, if x is in S, then qx is also in S because

$$A(qx) = qAx = q.0 = 0$$
$$\Rightarrow \quad A(qx) = 0$$

meaning that qx is in S, whenever x is.

Also, for every x in S, $-x$ is in S because

$$A(-x) = -Ax = -0 = 0$$
$$\Rightarrow \quad A(-x) = 0$$

meaning that $-x$ is in S, whenever x is. Finally, the zero vector $\mathbf{0}$ is in S because $A\mathbf{0} = \mathbf{0}.$ Thus, the solution space of the linear homogeneous system is a vector space in the case of non uniqueness.

Now we assume uniqueness of solution to $AX = \mathbf{0}$; that is, the zero vector only. It is easy to see that the zero vector trivially satisfies all the axioms of any vector space – meaning that the zero vector alone can stand on its own as a vector space. Try to convince yourself that the zero vector actually does satisfy all the axioms. The zero vector space V, is usually represented as V = {0}. Any vector space V not the zero vector space is called a **nonzero** vector space.

The conclusion is this: whichever way, the solution space of the linear homogeneous system $AX = \mathbf{0}$, is a real vector space.

SUBSPACES AND SPANNING

One obvious corollary of the axioms of a vector space V is that for every set of vectors $\mathbf{v}_1, \mathbf{v}_2, \mathbf{v}_3, \dots \mathbf{v}_n$ in V, any linear combination of this set of vectors is also in V; that is, any vector $\mathbf{v}$ of the form

$$\mathbf{v} = \mathbf{v}_1 + \mathbf{v}_2 + \mathbf{v}_3 + \cdots \mathbf{v}_n \text{ is in V.}$$

In fact any vector in V can always be written as a linear combination of v_i by choosing the right set of scalars a_i , $i \in \{1, 2, \ldots n\}$. The solution space of a homogeneous linear system illustrates this claim. Let $x_1, x_2, \ldots x_n$ be elements of the solution space S of the homogeneous linear system

$$AX = 0, \text{ then}$$

$$Ax_i = 0, \text{ for each } i \in \{1, 2, \ldots n\}.$$

$$\sum_{i=0}^{n} Ax_i = 0, \text{ since each } Ax_i = 0$$

$$\Rightarrow \quad A \sum_{i=1}^{n} x_i = 0, \text{ since } A \text{ is independent of } i,$$

meaning that

$$\sum_{i=1}^{n} x_i = x_1 + x_2 + \ldots x_n \text{ is in S.}$$

Moreover, the vector $v = a_1 x_1 + a_2 x_2 + \cdots a_n x_n$ is in S, because

$$A \left(\sum_{i=1}^{n} a_i x_i \right) = \sum_{i=1}^{n} A a_i x_i = 0, \text{ since each } A a_i x_i = a_i(Ax_i) = a_i.0 = 0$$

This means that

$$\sum_{i=1}^{n} a_i x_i \quad \text{ is in S.}$$

The set Q of vectors $v_1, v_2, v_3, \ldots v_n$ of a vector space V is said to **span** V if every vector in V can be written as a linear combination of $v_1, v_2, v_3, \ldots v_n$. That is, every vector v in V is of the form

$$v = a_1 v_1 + a_2 v_2 + \cdots a_n v_n , \text{ for certain scalars } a_1, a_2, \ldots a_n.$$

For example, the vectors $(1,0), (0,1)$ span $\mathbb{R}^2$; every vector in $\mathbb{R}^2$ can be written in the form $a_1 (1,0) + a_2 (0,1) = (a_1,0) + (0,a_2) = (a_1,a_2)$; also, every vector v in $\mathbb{R}^3$ can be written as

$$v = a_1 (1,0,0) + a_2(0,1,0) + a_3(0,0,1)$$

$$= (a_1, 0, 0) + (0, a_2, 0) + (0, 0, a_3) = (a_1, a_2, a_3).$$

Therefore, we say that the set $Q = \{(1,0), (0,1)\}$ spans $\mathbb{R}^2$, and the set $Q = \{(1,0,0), (0,1,0), (0,0,1)\}$ spans $\mathbb{R}^3$.

Example 2.5

Does the set of vectors

$S = \{1,0,1,0), (-1,0,1,0), (2,0,3,1), (-1,0,1,2)\}$ Span $\mathbb{R}^4$?

Solution

NO: This is because each of the vectors in S has zero as its second coordinate; thus, any linear combination of these vectors must also have zero as its second number. This means that vectors in $\mathbb{R}^4$ whose second coordinate values are not zero cannot be written as a linear combination of the elements of set S. This implies that the set S does not span $\mathbb{R}^4$.

SUBSPACES

Referring to the example above, it is easily seen that, although S does not span $\mathbb{R}^4$, the set of all vectors in $\mathbb{R}^4$ of the form $(a_1, 0, a_3, a_4)$ is spaned by S. Moreover, this set is a vector space in its own right as all the axioms for a vector space are satisfied by this set. This vector space which is a subset of $\mathbb{R}^4$ is called a **subspace** of $\mathbb{R}^4$. Let the subspace be W, and let the vectors $(a_1, 0, a_3, a_4)$ and $(a_1', 0, a_3', a_4')$ be in W, then $(a_1 + a_1', 0, a_3 + a_3', a_4 + a_4')$ is also in W.

For every vector $v = (a_1, 0, a_3, a_4)$ in W, there is a vector v_1 in W such that $v_1 = pv$ for any scalar p. This is because

$$v_1 = p(a_1, 0, a_3, a_4) = (pa_1, 0, pa_3, pa_4), \text{ which is in W.}$$

For every $v = (a_1, 0, a_3, a_4)$ in W, $-v = (-a_1, 0, -a_3, -a_4)$ is in W.

It is clear that the vector $(0,0,0,0)$ is in W. The reader should be convinced that the other axioms are also satisfied by W. Therefore, W is a vector space. Since W is a subset of $\mathbb{R}^4$, then W is said to be a subspace of $\mathbb{R}^4$. Note that every vector space is a subspace of itself since every set is a subset of itself.

In summary, if V is a vector space and W is a nonempty subset of V, then W is a subspace of V if and only if the following two conditions hold

i)   *for every w_1 and w_2 in* W, $w_1 + w_2$ *is also in* W

ii)   *for every w in* W *and any scalar p, pw is also in* W.

Example 2.6

Let V be the Cartesian 3 – space i.e. R^3 and let W be the set of all vectors of the form (x_1, x_2, x_3) such that $x_1 + x_2 + x_3 = 0$.
Show that W is a subspace of R^3.

Solution

We only need to show the above two properties of a subspace of a vector space.
First, the zero vector $(0,0,0)$ is in W because $0 + 0 + 0 = 0$; thus, W is not empty.

$$Let\ w_1 = (x_1, x_2, x_3)\ ; x_1 + x_2 + x_3 = 0\ and$$

$$w_2 = (x_1', x_2', x_3')\ ; x_1' + x_2' + x_3' = 0\ \text{be in}\ W, \text{then}$$

$$w_1 + w_2 = (x_1, x_2, x_3) + (x_1', x_2', x_3') = (x_1 + x_1', x_2 + x_2', x_3 + x_3')\ ; x_1 + x_2 + x_3$$
$$= 0\ and\ x_1' + x_2' + x_3' = 0$$

The vector $w_1 + w_2$ is in W because

$$(x_1 + x_1') + (x_2 + x_2') + (x_3 + x_3') = x_1 + x_2 + x_3 + x_1' + x_2' + x_3' = 0 + 0 = 0$$

Again, for every $w = (x_1, x_2, x_3)$ in W, and any scalar p, pw is in W because
$pw = p(x_1, x_2, x_3) = (px_1, px_2, px_3)$. Since $x_1 + x_2 + x_3 = 0$, then

$$p(x_1 + x_2 + x_3) = px_1 + px_2 + px_3 = 0.$$

This implies that pw is in W. Thus, W is a subspace of R^3.

It is not difficult to verify that all the vectors in W above can be written as a linear combination of the vectors $w_1 = (-1,0,1)$ and $w_2 = (0,1,-1)$. That is, any vector

$w = (x_1, x_2, x_3)$; $x_1 + x_2 + x_3 = 0$ in W can be written as

$$(x_1, x_2, x_3) = p_1(-1,0,1) + p_2(0,1,-1)$$
$$= (-p_1, 0, p_1) + (0, p_2, -p_2)$$
$$= (-p_1, p_2, p_1 - p_2), \text{ which clearly satisfies}$$

the condition that

$$x_1 + x_2 + x_3 = 0.$$

Example 2.7

If $S = \{v_1, v_2, \dots v_n\}$ is any collection of vectors from the vector space V, show that the set of all linear combinations

$v = p_1 v_1 + p_2 v_2 + \cdots p_n v_n$ (for scalars $p_1, p_2, \dots p_n$) of these vectors form a vector space; hence, a subspace of V.

Solution

Let $L(S)$ be this subspace.

For every $v = p_1 v_1 + p_2 v_2 + \cdots p_n v_n$ in $L(S)$ and any scalar α, the vector

$\alpha v = \alpha p_1 v_1 + \alpha p_2 v_2 + \cdots \alpha p_n v_n$ is clearly in $L(S)$; also, for

$v = p_1 v_1 + p_2 v_2 + \cdots p_n v_n$ and $u = q_1 v_1 + q v_2 + \cdots q_n v_n$ in $L(S)$, the

vector $u + v = (q_1 + p_1)v_1 + (q_2 + p_2)v_2 + \cdots (q_n + p_n)v_n$ is clearly in $L(S)$.

Therefore, $L(S)$, being a subset of V, is a subspace of V having satisfied the above properties. The subspace $L(S)$ is referred to as the **linear span** of the set

$S = \{v_1, v_2, \dots v_n\}$. The vector set $S = \{v_1, v_2, \dots v_n\}$ is referred to as the **spanning set** for $L(S)$.

APPLICATIONS OF LINEAR ALGEBRA

BASES AND DIMENSIONS

It is not difficult to see that the sets $S_1 = \{(1,0), (0,1), (1,1)\}$,

$S_2 = \{(1,-1), (1,0), (0,-1)\}$, and $S_3 = \{(1,0), (0,1)\}$ each span R^2, but S_1 and S_2 are each linearly dependent sets, while S_3 is linearly independent. Note also that no set S containing only one vector spans R^2. These observations raise the question: for a vector space V, is there a minimal spanning set? That is, is there a spanning set for V (or simply a spanner of V) that contains a minimum number of element (vectors)? Will there always be a linearly independent spanner of V? The answer to these questions is yes. This leads us to the concepts of bases and dimension.

BASES

Let V be a nonzero vector space. The set of vector $S = \{v_1, v_2, \dots v_n\}$ is called a **basis** for (or of) V if each of the following hold.

i) $L(S) = V$; *that is, the linear span of S is V or simply the set S spans* V.

ii)   *The set S is linearly independent.*

Thus, we can define a basis for a vector space V as a linearly independent spanner (or spanning set) of V. For example, the set $S = \{(1,0), (0,1), (1,1)\}$ is a spanning set (or spanner) of V, but is linearly dependent; therefore, S is not a basis for R^2.

The set $S = \{(1,0), (0,1)\}$ clearly spans R^2; moreover, this set is linearly independent. Therefore, the set $S = \{(1,0), (0,1)\}$ is a basis for R^2. Again, note that the sets $S_1 = \{(1,1), (1,-1)\}$, $S_2 = \{(-1,-1), (1,0)\}$ are also bases for R^2. This means that the basis for a vector space is **not unique**. However, one thing is common to all bases for a given vector space: the number of elements (vectors) in each base is always the same. That is, the cardinality of the bases for a vector space is invariant.

The maximum number of linearly independent vectors that span a vector space is called the **dimension** of the vector space. That is,

i) *If V is a vector space and $S = \{v_1, v_2, \ldots v_n\}$ is a basis for V, then every set containing more than n vectors is linearly dependent, and hence cannot be a basis for V.*

ii) *If $S_1 = \{v_1, v_2, \ldots v_n\}$ and $S_2 = \{u_1, u_2, \ldots u_m\}$ are two bases for V, then $n = m$. This number $n = m$ is called the dimension of V.*

iii) *The vector space R^n has a dimension of n; that is, the Cartesian n-space is n-dimensional.*

iv) *If V is an n-dimensional vector space and $S = \{v_1, v_2, \ldots v_n\}$ is a linearly independent set, then S is a basis for V.*

v) *If V is an n-dimensional vector space, and $S = \{v_1, v_2, \ldots v_n\}$ spans V, then S is a basis for V.*

The proofs are left to the reader as exercises

ROW SPACE AND COLUMN SPACE

The terms **row space** and **column space** relate to the row vectors and column vectors of a matrix. In particular, the row space of an nxm matrix A with real entries a_{ij} is defined as the subspace of R^n spanned by the n row- vectors of A; that is, the row space of a matrix A is the linear span of the row vectors of A. Similarly, the column space of A is the linear span of the column vectors of A; that is, the subspace of R^m spanned by the column vectors of A. The dimension of the row space of A is called **row rank** of A, while the dimension of the column space of A is called **column rank** of A. In other words, the row rank of A is the cardinality of the basis of the row space of A, which is the same thing as the *minimum number* of the row vectors of A that span the row space of A or the *maximum number* of linearly independent row vectors of A that span the row space of A. Similar definitions hold for the column rank of A. This means that to determine the row rank of a matrix A, we carry out elementary row operations and reduce A to its row-echelon form. The number of non zero rows in the row-echelon form is the row rank of A.

The column rank of A is determined by carrying out elementary row operations on A^T until it is reduced to its row-echelon form. The number of non zero rows of this echelon form gives us the column rank of A.

Example 2.8

Find the row space, row rank, column space, and column rank of the matrix

$$A = \begin{pmatrix} 1 & -1 & 0 \\ 2 & 3 & 1 \\ 3 & -2 & 4 \end{pmatrix}.$$

Solution

The row space is the subspace of R^3 spanned by the vectors

$(-1, -1, 0)$, $(2, 3, 1)$ *and* $(3, -2, 4)$. This means that the row space is the subspace of R^3 whose elements are of the form

$v = p_1(-1, -1, 0) + p_2(2, 3, 4) + p_3(3, -2, 4)$ for scalars p_1, p_2, p_3. The row space is also the linear span of the row vectors. The column space of A is the linear span of the column vectors $\begin{pmatrix} 1 \\ 2 \\ 3 \end{pmatrix}, \begin{pmatrix} -1 \\ 3 \\ -2 \end{pmatrix}$, and $\begin{pmatrix} 0 \\ 1 \\ 4 \end{pmatrix}$; that is, the set of all vectors of

the form

$$v = q_1 \begin{pmatrix} 1 \\ 2 \\ 3 \end{pmatrix} + q_2 \begin{pmatrix} -1 \\ 3 \\ -2 \end{pmatrix} + q_3 \begin{pmatrix} 0 \\ 1 \\ 4 \end{pmatrix} \text{ for real } q, q_2, q_3.$$

Now for row rank, we have

$$\begin{pmatrix} 1 & -1 & 0 \\ 2 & 3 & 1 \\ 3 & -2 & 4 \end{pmatrix} (R_2\text{-}2R_1),(R_3\text{-}3R_1) \begin{pmatrix} 1 & -1 & 0 \\ 0 & 5 & 1 \\ 0 & 1 & 4 \end{pmatrix} (R_2\text{-}5R_3),(R_2\sim R_3) \begin{pmatrix} 1 & -1 & 0 \\ 0 & 0 & -19 \\ 0 & 1 & 4 \end{pmatrix}.$$

Therefore, the row rank of A is 3 since there are three nonzero rows in its row-echelon form.

For column rank, we perform elementary row operations on A^T as explained earlier

$$A^T = \begin{pmatrix} 1 & 2 & 3 \\ -1 & 3 & -2 \\ 0 & 1 & 4 \end{pmatrix} (R_2+R_1) \begin{pmatrix} 1 & 2 & 3 \\ 0 & 5 & 2 \\ 0 & 1 & 4 \end{pmatrix} (R_2\sim R_3), (R_3\text{-}5R_2) \begin{pmatrix} 1 & 2 & 3 \\ 0 & 1 & 4 \\ 0 & 0 & -19 \end{pmatrix}.$$

Therefore the column rank of A is 3.

It is not an accident that the row-rank of the matrix A is equal to its column rank. Recall that the problem of finding rank is equivalent to that of finding the maximum number of linearly independent vectors. Since the row vectors of a matrix are linearly independent if and only if the column vectors are linearly independent; therefore, it is always the case that row rank of A is always equal to the column rank of A.

i)   *The row vectors of a matrix are linearly independent if and only if its column vectors are linearly independent*

ii)   *The row vectors of a matrix are linearly dependent if and only if its column vectors are linearly dependent.*

iii)   *The column rank of a matrix is always equal its row rank.*

iv)   *Elementary row operations on a matrix do not change its row-space.*

This means that the row space of matrix A above can also be taken as a subspace of $\mathbb{R}^3$ spanned by the vectors $(1, -1, 0), (0, 0, -19), (0, 1, 4)$.

v)   *The nonzero row vectors of the echelon form of a matrix form a basis of its row space.*

vi)   *The nonzero column vectors of the transposition of the row- echelon form of the transposition a matrix form a basis for its column space.*

Example 2.9

Find the row rank of the matrix

$$A = \begin{pmatrix} 1 & 0 & 1 & 2 \\ 2 & 1 & 0 & 3 \\ 1 & -1 & 3 & 3 \end{pmatrix}.$$

Solution

$$\begin{pmatrix} 1 & 0 & 1 & 2 \\ 2 & 1 & 0 & 3 \\ 1 & -1 & 3 & 3 \end{pmatrix} \underset{\sim}{(R_2-2R_1)} \begin{pmatrix} 1 & 0 & 1 & 2 \\ 0 & 1 & -2 & -1 \\ 1 & -1 & 3 & 3 \end{pmatrix} \underset{\sim}{(R_3+R_2),(R_3-R_1)} \begin{pmatrix} 1 & 0 & 1 & 2 \\ 0 & 1 & -2 & -1 \\ 0 & 0 & 0 & 0 \end{pmatrix}$$

Therefore, the row rank of $A = 2$; also, the basis for the row- space of A is the set $S = \{(1,0,1,2), (0,1,-2,-1)\}$.

Example

Find a basis for the space spanned by the set of vectors
$S = \{(3, 1, 0), (0, 4, 2), (-1, 5, 3), (1, 1, -4)\}.$

Solution

$$
\begin{pmatrix} 3 & 1 & 0 \\ 0 & 4 & 2 \\ -1 & 5 & 3 \\ 1 & 1 & -4 \end{pmatrix}
(R_1 \sim R_4), (R_4 - 3R_1), (R_3 + R_1)
\begin{pmatrix} 1 & 1 & -4 \\ 0 & 4 & 2 \\ 0 & 6 & -1 \\ 0 & -2 & 12 \end{pmatrix}
$$

$$
(\tfrac{1}{2}R_2), (R_3 - 3R_2), (R_4 + R_2)
\begin{pmatrix} 1 & 1 & -4 \\ 0 & 2 & 1 \\ 0 & 0 & -4 \\ 0 & 0 & 13 \end{pmatrix}
(\tfrac{1}{4}R_3), (R_4 + 13R_3)
\begin{pmatrix} 1 & 1 & -4 \\ 0 & 2 & 1 \\ 0 & 0 & -1 \\ 0 & 0 & 0 \end{pmatrix}.
$$

Therefore the basis for the linear span of the set S above is

$S_1 = \{(1, 1, -4), (0, 2, 1), (0, 0, -1)\}$. Therefore the linear span of S is equal to

$$
L(S) = L(S_1) = \mathbb{R}^3.
$$

Example 2.10

Find a basis for the column space of the matrix

$$
A = \begin{pmatrix} -1 & 0 & 1 & 1 \\ 3 & -1 & 2 & 0 \\ 1 & 3 & -1 & 4 \end{pmatrix}.
$$

Solution

The basis for the column- space of A is equal to the basis for the row- space of A^T with each row vector transposed.

$$
A^T = \begin{pmatrix} -1 & 3 & 1 \\ 0 & -1 & 3 \\ 1 & 2 & -1 \\ 1 & 0 & 4 \end{pmatrix}
(R_3 + R_1), (R_4 + R_1)
\begin{pmatrix} -1 & 3 & 1 \\ 0 & -1 & 3 \\ 0 & 5 & 0 \\ 0 & 3 & 5 \end{pmatrix}
$$

$$(R_3 + 5R_2), (R_4 + 3R_2) \begin{pmatrix} -1 & 3 & 1 \\ 0 & -1 & 3 \\ 0 & 0 & 15 \\ 0 & 0 & 20 \end{pmatrix} \left(\tfrac{1}{15} R_3\right) (R_4 - 20R_3) \begin{pmatrix} -1 & 3 & 1 \\ 0 & -1 & 3 \\ 0 & 0 & 1 \\ 0 & 0 & 0 \end{pmatrix}$$

Therefore, the basis for the column-space of A is the set

$$S_1 = \left\{ \begin{pmatrix} -1 \\ 3 \\ 1 \end{pmatrix}, \begin{pmatrix} 0 \\ -1 \\ 3 \end{pmatrix}, \begin{pmatrix} 0 \\ 0 \\ 1 \end{pmatrix} \right\}; \text{ also, the column rank is 3.}$$

Check that the row rank is also 3; it is not an accident. The column rank of a matrix is always equal to its row rank as previously mentioned.

REAL INNER PRODUCT SPACES

We have introduced the concept of real Inner Product Spaces in earlier work. Now we extend this idea to an arbitrary vector space. This is because the inner product treated is just one out of several inner products. The main importance of imposing an inner product on a vector space is to establish a measure of distance between any two elements of the vector space, the length (norm) of an element, and also the "angle" between two elements. Any vector space with a defined inner product is called **inner product space**. If the field of scalars over which the space is distributed is the set of real numbers, then such an inner product space is called **real inner product space.** Any function that satisfies the following properties is necessarily an inner product function.

Let V be any real vector space. Suppose to each pair u and v of vectors in V we can assign a real number $\langle u, v \rangle$ with the following properties. For all u, v, w in V and real numbers p_1, p_2

i) $\langle u, v \rangle = \langle v, u \rangle$ $--$ symmetric property or simply symmetry.

ii) $\langle p_1 u + p_2 v, w \rangle = p_1 \langle u, w \rangle + p_2 \langle v, w \rangle$

iii) $\langle u, u \rangle \geq 0$

iv) $\langle u, u \rangle = 0$ if and only if $u = 0$

This function $\langle u, v \rangle$ defined on v is called an **inner product** in V, and V is an inner product space with respect to this defined inner product $\langle u, v \rangle$.

Example 2.11

If the vector space V is a Cartesian $-n$ space, R^n, and suppose that the vectors $v = (v_1, v_2, v_3, \ldots v_n)$, and $u = (u_1, u_2, u_3, \ldots u_n)$ are in V where all v_i, u_i, are real. Show that the function $\langle v, u \rangle$ defined by

$\langle v, u \rangle = v_1 u_1 + v_2 u_2 + \cdots v_n u_n$ is an inner product in R^n.

Solution

We merely need to show that the function $\langle v, u \rangle$, satisfies the four properties of an inner product.

i) $\langle u, v \rangle = u_1 v_1 + u_2 v_2 + \cdots u_n v_n = v_1 u_1 + v_2 u_2 + \cdots v_n u_n = \langle v, u \rangle.$

$\Rightarrow \langle u, v \rangle = \langle v, u \rangle$; symmetry is established.

ii)   Let $w = (w_1, w_2, w_3, \ldots w_n)$, then for scalars p_1, p_2

$p_1 u = (p_1 u_1, p_1 u_2, p_1 u_3, \ldots p_1 u_n)$, $p_2 v = (p_2 v_1, p_2 v_2, p_2 v_3, \ldots p_2 v_n)$

$$p_1 u + p_2 v = (p_1 u_1 + p_2 v_1, p_1 u_2 + p_2 v_2, \ldots p_1 u_n + p_2 v_n)$$

$$
\begin{aligned}
\langle p_1 u + p_2 v, w \rangle &= (p_1 u_1 + p_2 v_1) w_1 + (p_1 u_2 + p_2 v_2) w_2 + \cdots (p_1 u_n + p_2 v_n) w_n \\
&= p_1 u_1 w_1 + p_2 v_1 w_1 + p_1 u_2 w_2 + p_2 v_2 w_2 + \cdots p_1 u_n w_n + p_2 v_n w_n \\
&= p_1 u_1 w_1 + p_1 u_2 w_2 + \cdots p_1 u_n w_n + p_2 v_1 w_1 + p_2 v_2 w_2 + \cdots p_2 v_n w_n \\
&= p_1 (u_1 w_1 + u_2 w_2 + \cdots u_n w_n) + p_2 (v_1 w_1 + v_2 w_2 + \cdots v_n w_n) \\
&= p_1 \langle u, w \rangle + p_2 \langle v, w \rangle
\end{aligned}
$$

$$\langle u, u \rangle = u_1 u_1 + u_2 u_2 + \cdots u_n u_n = u_1^2 + u_2^2 + \cdots u_n^2$$

$$\Rightarrow \langle u, u \rangle \geq 0 \text{ , since each } u_i^2 \geq 0.$$

Property (iv) easily follows from property (iii) when $u = 0$. Therefore, the function $\langle v, u \rangle$, as defined, is an inner product in $\mathbb{R}^n$. This particular inner product is called the **standard inner product.** It is also called the **Euclidean inner product.** Therefore, the Cartesian n- space $\mathbb{R}^n$ with respect to the standard (Euclidean) inner product is called **Euclidean n-space.**

Example 2.12

Show that the function $\langle u, v \rangle = 2u_1v_1 + 3u_2v_2$, where v and u are as previously defined, is an inner product on $\mathbb{R}^2$.

Solution

i) $\langle v, u \rangle = 2v_1u_1 + 3v_2u_2 = 2u_1v_1 + 3u_2v_2 = \langle u, v \rangle$

       $\langle v, u \rangle = \langle u, v \rangle$ $-$ symmetry.

ii)   For scalars $p_1, p_2,$ and $w = (w_1, w_2)$,

$$\langle p_1v + p_2u, w \rangle = \langle (p_1v_1, p_1v_2) + (p_2u_1, p_2u_2), (w_1, w_2) \rangle$$
$$= \langle (p_1v_1 + p_2u_1, p_1v_2 + p_2u_2), (w_1, w_2) \rangle$$
$$= 2(p_1v_1 + p_2u_1)w_1 + 3(p_1v_2 + p_2u_2)w_2$$
$$= 2p_1v_1w_1 + 3p_1v_2w_2 + 2p_2u_1w_1 + 3p_2u_2w_2$$
$$= p_1(2v_1w_1 + 3v_2w_2) + p_2(2u_1w_1 + 3u_2w_2)$$
$$= p_1\langle v, w \rangle + p_2\langle u, w \rangle$$
$$\langle u, u \rangle = 2u_1u_1 + 3u_2u_2 = 2u_1^2 + 3u_2^2 \geq 0$$
$$When\ u = 0 = (0,0)\ , \langle u, u \rangle = 0.$$

Therefore, the function $\langle u, v \rangle = 2u_1v_1 + 3u_2v_2$ defines an inner product on $\mathbb{R}^2$. However, this inner product is not the standard inner product.

Example 2.13

Let M_n be the space whose elements are real nxn matrices. If A and B are in M_n, show that the function $\langle A, B \rangle$ defined by $\langle A, B \rangle = tr(A^T B)$ is a real inner product, where "tr" means "trace".

The trace of a matrix $D(d_{ij})$ is defined as the sum of all the elements (entries) on the main diagonal. That is,

$$tr(D) = \sum_{i=1}^{n} d_{ii} \quad \text{where } D \text{ is an } nxn \text{ matrix.}$$

Solution

First note that the diagonal elements of any nxn matrix, upon transposing, do not change. This simply tells us that, if A is any nxn matrix, then

$$tr(A) = tr(A^T).$$

Recall also that for any two matrices A and B compatible for multiplication,

$$(AB)^T = B^T A^T$$
$$\Rightarrow (B^T A)^T = A^T (B^T)^T = A^T B \text{ , } since \ (B^T)^T = B$$
$$tr(A^T B) = tr((B^T A)^T) = tr(B^T A)$$
$$\Rightarrow \langle A, B \rangle = \langle B, A \rangle \qquad \qquad ---symmetry.$$

i)   If matrix $C(nxn)$ is also in M_n, suppose that p_1, p_2 are any real numbers, then

$$\langle p_1 A + p_2 B, C \rangle = tr\big(C^T(p_1 A + p_2 B)\big), \quad due \ to \ symmetry.$$
$$= tr(C^T p_1 A + C^T p_2 B)$$
$$= tr(p_1 C^T A + p_2 C^T B)$$

Be informed that $tr(A + B) = tr(A) + tr(B)$; this is empirical.

$$\Rightarrow \quad \langle p_1 A + p_2 B, C \rangle = tr(p_1 C^T A) + tr(p_2 C^T B).$$

Again, convince yourself that, for any scalar $p, tr(pA) = p \, tr(A)$.

$$\Rightarrow \quad \langle p_1 A + p_2 B, C \rangle = p_1 tr(C^T A) + p_2 tr(C^T B)$$
$$= p_1 \langle A, C \rangle + p_2 \langle B, C \rangle$$

ii)

$$\langle A, A \rangle = tr(A^T A) = \sum_{i=1}^{n} \sum_{j=1}^{n} (a_{ij})^2 \geq 0.$$

iii)   When A is a null matrix, each entry of A is zero. Therefore,

$$\langle A, A \rangle = tr(A^T A) = \sum_{i=1}^{n} \sum_{j=1}^{n} (0)^2 = 0.$$

Therefore the function $\langle A, B \rangle = tr(A^T B)$ defines an inner product on M_n.

Example 2.14

Let P_n be the space of real polynomials of degree n or less. If for any two polynomials $q(x), s(x)$ in P_n, the function $\langle q(x), s(x) \rangle$ is defined as $\langle q(x), s(x) \rangle = \int_0^1 q(x)s(x)dx$. Show that $\langle q(x), s(x) \rangle$ is an inner product on P_n.

Solution

i)

$$\langle q(x), s(x) \rangle = \int_0^1 q(x)s(x)dx = \int_0^1 q(x)s(x)dx$$
$$\Rightarrow \langle q(x), s(x) \rangle = \langle s(x), q(x) \rangle$$

ii)   If $t(x)$ is also any function in P_n, for any scalars k_1 and k_2,

$$\langle k_1 q(x) + k_2 s(x), t(x) \rangle = \int_0^1 (k_1 q(x) + k_2 s(x))t(x)dx$$

$$= \int_0^1 \left(k_1 q(x)t(x) + k_2 s(x)t(x) \right) dx$$

$$= \int_0^1 k_1 q(x)t(x)dx + \int_0^1 k_2 s(x)t(x)dx; \text{linearity of the integral operator}$$

$$= k_1 \int_0^1 q(x)t(x)dx + k_2 \int_0^1 s(x)t(x)dx$$

$$= k_1 \langle q(x), t(x) \rangle + k_2 \langle s(x), t(x) \rangle$$

iii

$$\langle q(x), q(x) \rangle = \int_0^1 q(x)q(x)dx$$

$$= \int_0^1 (q(x))^2 dx \text{ , which is always nonnegative}$$

$$\Rightarrow \quad \langle q(x), q(x) \rangle \geq 0.$$

iv)   When $q(x) = 0$, the zero polynomial, we have

$$\langle q(x), q(x) \rangle = \int_0^1 0^2 \, dx = 0$$

Therefore, $\langle q(x), q(x) \rangle$, as defined, is an inner product on P_n.

Example 2.15

Let P_n, again, be the space of real polynomials of degree n or less. If for any two polynomials $q(x), s(x)$, in P_n, the function $\langle q(x), s(x) \rangle$ is defined by
$\langle q(x), s(x) \rangle = q_0 s_0 + q_1 s_1 + q_2 s_2 + \cdots q_n s_n$, where

$$q(x) = q_0 + q_1 x + q_2 x^2 + \cdots q_n x^n$$
$$s(x) = s_0 + s_1 x + s_2 x^2 + \cdots s_n x^n$$

Show that $\langle q(x), s(x) \rangle$ is an inner product on P_n. Moreover, this inner product resembles the Euclidean inner product on $\mathbb{R}^n$.

Solution

$\langle s(x), q(x) \rangle = s_0 q_0 + s_1 q_1 + s_2 q_2 + \cdots s_n q_n$, which is clearly the same as $\langle q(x), s(x) \rangle$. Therefore, $\langle s(x), q(x) \rangle = \langle q(x), s(x) \rangle$.
Let $w(x) = w_0 + w_1 x + w_2 x^2 + \cdots w_n x^n$ be any other polynomial in P_n, then for any scalars k_1 and k_2 ,

$\langle k_1 q(x) + k_2 s(x), w(x) \rangle$

$= \langle k_1 q_0 + k_1 q_1 x + k_1 q_2 x^2 + \cdots k_1 q_n x^n + k_2 s_0 + k_2 s_1 x + k_2 s_2 x^2 + \cdots k_2 s_n x^n, w_0 + w_1 x + w_2 x^2 + \cdots w_n x^n \rangle$

$= \langle k_1 q_0 + k_2 s_0 + (k_1 q_1 + k_2 s_1)x + (k_1 q_2 + k_2 s_2)x^2 + \cdots (k_1 q_n + k_2 s_n)x^n, w_0 + w_1 x + w_2 x^2 + \cdots w_n x^n \rangle$

$= (k_1 q_0 + k_2 s_0)w_0 + (k_1 q_1 + k_2 s_1)w_1 + (k_1 q_2 + k_2 s_2)w_2 + \cdots (k_1 q_n + k_2 s_n)w_n$

$= k_1(q_0 w_0 + q_1 w_1 + q_2 w_2 + \cdots q_n w_n) + k_2(s_0 w_0 + s_1 w_1 + s_2 w_2 + \cdots s_n w_n)$

$= k_1 \langle q(x), w(x) \rangle + k_2 \langle s(x), w(x) \rangle$

iii) $\quad \langle q(x), q(x) \rangle = q_0 q_0 + q_1 q_1 + q_2 q_2 + \cdots q_n q_n$

$$= q_0^2 + q_1^2 + q_2^2 + \cdots q_n^2 \geq 0$$

iv) When $q(x) = 0$, the zero polynomial

$$\langle q(x), q(x) \rangle = 0^2 + 0^2 + 0^2 + \cdots = 0$$

Therefore $\langle q(x), s(x) \rangle = q_0 s_0 + q_1 s_1 + q_2 s_2 + \cdots q_n s_n$ defines an inner product on P_n. This inner product resembles the Euclidean inner product on R^n.

LENGTH AND ANGLE IN REAL INNER PRODUCT SPACE

Now we have enough tools to generalize the concept of distance in R^2 and R^3. Let V be any real inner product space, with the inner product function (u, v) for any u, v in V. We define the **length** or **norm** of an element v in V as

$$\|v\| = \langle v, v \rangle^{\frac{1}{2}} \qquad - - - - - P165$$

Note that this new notation for norm (length) does not portray anything different from the conventional notation for the magnitude of a vector $|v|$. That is,

$$\|v\| = |v| = v$$

Henceforth, we adopt the notation $|v|$ for norm of v.

Example 2.16:

Let $V = R^4$ with inner product $\langle u, v \rangle$ defined by

$$\langle u, v \rangle = 2u_1 v_1 + u_2 v_2 + 3u_3 v_3 + u_4 v_4.$$

Find the norm (length) of

$$w = (1, 2, -1, 5)$$

Solution

$$\text{Norm of } w = |w| = \langle w, w \rangle^{\frac{1}{2}}$$

$$\Rightarrow \quad |w| = \left(2w_1^2 + w_2^2 + 3w_3^2 + w_4^2\right)^{\frac{1}{2}}$$

$$= (2.1^2 + 2^2 + 3.(-1)^2 + 5^2)$$

$$= \sqrt{34}$$

Example 2.17

Let P_n be the inner product space of Polynomials of degree n or less with the integral Inner Product. Find the norm of a Polynomial $p(x) = e^x$.

Solution

$$|p(x)| = \langle p(x), p(x) \rangle^{\frac{1}{2}}$$

$$= \left(\int_0^1 p(x)\, p(x)\, dx\right)^{\frac{1}{2}}$$

$$= \left(\int_0^1 (e^x)^2\, dx\right)^{\frac{1}{2}} = -\left(\int_0^1 e^{2x}\, dx\right)^{\frac{1}{2}}$$

$$\Rightarrow \quad |p(x)| = \left(\left|\frac{1}{2}e^{2x}\right|_0^1\right)^{\frac{1}{2}}$$

$$= \frac{(e^2 - 1)^{\frac{1}{2}}}{\sqrt{2}}$$

Cauchy schwarz inequality for an arbitrary inner product space

Let V be an inner product space, for all vectors u and v in V, we have

$$|\langle v, v \rangle| \leq |u||v| \qquad\qquad ----- P166$$

96

Proof

Let

$$w = wu + v$$

$$\langle w, w \rangle = \langle wu + v, wu + v \rangle$$

$$= \langle wu, wu + v \rangle + \langle v, wu + v \rangle$$

$$= \langle wu, wu \rangle + \langle wu, v \rangle + \langle v, wu \rangle + \langle v, v \rangle$$

$$= w^2 \langle u, u \rangle + 2w \langle u, v \rangle + \langle v, v \rangle$$

$$= \langle u, u \rangle \left[\left(w + \frac{\langle u, v \rangle}{\langle u, u \rangle} \right)^2 - \left(\frac{\langle u, v \rangle}{\langle u, u \rangle} \right)^2 \right] + \langle v, v \rangle$$

$$= \langle u, u \rangle \left(w + \frac{\langle u, v \rangle}{\langle u, u \rangle} \right)^2 - \frac{\langle u, v \rangle^2}{\langle u, u \rangle} + \langle v, v \rangle$$

But $\langle w, w \rangle \geq 0$

$$\therefore \langle u, u \rangle \left(w + \frac{\langle u, v \rangle}{\langle u, u \rangle} \right)^2 - \frac{\langle u, v \rangle^2}{\langle u, u \rangle} + \langle v, v \rangle \geq 0$$

This inequality holds for any w value. In particular, when

$$w = -\frac{\langle u, v \rangle}{\langle u, u \rangle}, \text{ then}$$

$$-\frac{\langle u, v \rangle^2}{\langle u, u \rangle} + \langle v, v \rangle \geq 0$$

$$\langle u, u \rangle \langle v, v \rangle \geq \langle u, v \rangle^2 \text{, since } \langle u, u \rangle \geq 0.$$

$$\Rightarrow \quad |\langle v, v \rangle| \leq |v||u|$$

Triangular inequality for an arbitrary inner product space

Let V be an inner product space as earlier defined, and let v, u be in V. Then,

$$|v + u| \leq |v| + |u| \qquad\qquad ----P167$$

Proof

$$|v + u|^2 = \langle v + u, v + u \rangle$$

$$= \langle v, v \rangle + 2 \langle u, v \rangle + \langle u, u \rangle$$

since $\langle u,v\rangle \leq |\langle u,v\rangle|$,we have

$$|v + u|^2 \leq \langle v,v\rangle + 2|\langle u,v\rangle| + \langle u,u\rangle$$

since $|\langle u,v\rangle| \leq |u||v|$, we have

$$|v + u|^2 \leq |v|^2 + 2|u||v| + |u|^2$$
$$\Rightarrow |v + u|^2 \leq (|v| + |u|)^2$$
$$\Rightarrow |v + u| \leq |v| + |u|$$

Just as the ideas of length and distance have been generalized for an arbitrary inner product space, the concept of angle can be generalized.

The Cauchy Schwatz inequality helps us to achieve this idea of imposing the concept of angle on a generalized inner product space.

By definition, the angle between any two elements u, v of an arbitrary inner Product space is

$$\cos\theta = \frac{\langle u,v\rangle}{|u|,|v|} \leq 1 \; ; \; 0 \leq \theta \leq \pi$$

Recall that if $u \perp v$, then $\langle u,v\rangle = 0$. To generalize the idea of perpendicularity of two elements of an inner product space, we rather use the word *orthogonality*. Any two elements u, v of an arbitrary inner product space are said to be orthogonal if $\langle u,v\rangle = 0$.

Example 2.18

Show that the polynomials $q(x) = x - 1, s(x) = 3x - 1$ in P_1 are orthogonal in P_1, where P_1 is the inner product space of polynomials of degree one or less with the integral inner product.

Solution

$$\langle q(x), s(x)\rangle = \int_0^1 q(x)\, s(x)dx$$

$$\langle q(x), s(x) \rangle = \int_0^1 (x-1)(3x-1)dx$$

$$= \int_0^1 (3x^2 - 4x + 1)dx$$

$$= |x^3 - 2x^2 + x|_0^1$$

$$= 1^3 - 2.1^2 + 1 - 0 = 0$$

$\therefore$ $q(x)$ and $s(x)$ are orthogonal under the integral inner product.

Example 2.19

Find the value(s) of t so that the polynomial $p(x) = 1 - tx^2$ is a unit vector in P_2 with integral inner product.

Solution

If $p(x)$ is a unit vector, then

$$|p(x)| = 1$$

$$\Rightarrow \quad \langle p(x), p(x) \rangle^{1/2} = 1$$

$$\Rightarrow \quad \left(\int_0^1 (p(x))^2 \, dx \right)^{1/2} = 1$$

$$\therefore \quad \left(\int_0^1 (1 - tx^2)^2 \, dx \right)^{1/2} = 1$$

$$\Rightarrow \quad (|1 - 2tx^2 + t^2 x^4|_0^1)^{\frac{1}{2}} = 1$$

$$\Rightarrow \quad [1 - 2t + t^2 - (1 - 0 + 0)]^{\frac{1}{2}} = 1$$

$$\Rightarrow \quad t^2 - 2t - 1 = 0$$

$$t = 1 \pm \sqrt{2}$$

This means that the polynomials $1 - (1 + \sqrt{2})x^2$ and $1 - (1 - \sqrt{2})x^2$ are each unit vectors in P_n.

Example 2.20

Find the value(s) of t so that the matrix

$A = \begin{pmatrix} -1 & t \\ 3 & 2 \end{pmatrix}$ is a unit element (vector) in M_2 with the trace inner product.

Solution

If A is a unit element, then

$$(\langle A, A \rangle)^{1/2} = 1$$

$$\Rightarrow \qquad [tr(A^T A)]^{1/2} = 1$$

$$\Rightarrow \quad tr(A^T A) = 1$$

Recall that for any $n \times n$ matrix $A(a_{ij})$,

$$tr(A^T A) = \sum_{i=1}^{n} \sum_{j=1}^{n} (a_{ij})^2$$

$$tr(A^T A) = (-1)^2 + t^2 + 3^2 + 2^2 = 1$$

$$\Rightarrow \qquad = t^2 + 13 = 0$$

$$t = \pm i\sqrt{13}$$

Since M_2 is assumed to be a real inner product space; therefore, there is no t for which the matrix $A = \begin{pmatrix} -1 & t \\ 3 & 2 \end{pmatrix}$ is a unit vector in M_2.

Example 2.21

Find the value of t such that the matrix $A = \begin{pmatrix} \frac{1}{2} & t \\ \frac{1}{3} & \frac{2}{5} \end{pmatrix}$ is a unit vector in M_2 with the trace inner product.

Solution

For $|A| = 1$, we have

$$(\langle A, A \rangle)^{1/2} = 1$$

$$\Rightarrow \langle A, A \rangle = 1$$

$$tr\,(A^T A) = 1$$

$$\Rightarrow \left(\frac{1}{2}\right)^2 + t^2 + \left(\frac{1}{3}\right)^2 + \left(\frac{2}{5}\right)^2 = 1$$

$$t^2 = 1 - \frac{1}{4} - \frac{1}{9} - \frac{4}{25}$$

$$t^2 = \frac{431}{900}$$

$$\Rightarrow \quad t = \pm \frac{\sqrt{431}}{30}$$

Therefore the matrices $\begin{pmatrix} \frac{1}{2} & \frac{\sqrt{431}}{30} \\ \frac{1}{3} & \frac{2}{5} \end{pmatrix}$ and $\begin{pmatrix} \frac{1}{2} & \frac{-\sqrt{431}}{30} \\ \frac{1}{3} & \frac{2}{5} \end{pmatrix}$ are unit vectors in M_2 under

the trace inner product.

Orthonormal bases

Let $S = \{v_1, v_2, \dots v_n\}$ be a basis for an n-dimensional real inner product space V; S is called **Orthonormal basis** for V if

i)   *Every vector $v_i (1 \leq i \leq n)$ in S is a unit vector; that is $|v_i| = 1$ for each i.*

ii)   *For any two vectors v_i and $v_j, i \neq j, (1 \leq i, j \leq n)$, in V,*

   $\langle v_i, v_j \rangle = 0.$

In other words, a basis S for V is an orthonormal basis if and only if each member (element) of the basis is a unit vector, and the vectors in S are pairwise orthogonal.

GRAM – SCHMIDT PROCESS

Calculations are made easier when we work with an orthonormal basis of an inner product space, and not just any basis. For this purpose, it becomes a worthwhile effort to search for an orthonormal basis for an inner product space. However, the search is not a random type; it is systematic. This is because any nonorthonormal basis can be converted, following a definite pattern, to an orthonormal basis. This process of converting a nonorthonormal basis into an orthonormal basis is called the **Gram-Schmidt Process.** To see the significance of the Gram-Schmidt Process, let's consider the following example.

Example 2.22

The set $S = \{(1, -1), (2, 3)\}$ is clearly a basis for Euclidean 2-space. This is because the set spans R^2, and it is also linearly independent. This, therefore, means that any vector v in R^2 can be written as a linear combination of the basis vectors. That is, if v is in R^2, then there are scalars p_1 and p_2 such that $v = p_1(1, -1) + p_2(2, 3)$. Find the scalars p_1 and p_2.

Solution

If $v = (v_1, v_2)$, then

$$v = p_1(1, -1) + p_2(2, 3)$$
$$\Rightarrow \quad p_1 + 2p_2 = v_1$$
$$p_1 + 3p_2 = v_2$$
$$p_2 = \frac{v_1 + v_2}{5}, \quad p_1 = \frac{3v_1 + v_2}{5}$$
$$\Rightarrow \quad v = \frac{3v_1 + 2v_2}{5}(1, -1) + \frac{v_1 + v_2}{5}(2, 3)$$

The tedium of the calculation we just did might not be felt in this 2-dimensional case. I can assure you that this same process we just completed becomes extremely tedious if the dimension of the space is significantly high. This is where the orthonormal basis comes in. To see this, let us consider the following example.

Example 2.23

Let the basis $S = \{v_1, v_2, \ldots v_n\}$ be such that $|v_i| = 1$ and $\langle v_i, v_j \rangle = 0$; $1 \leq i, j \leq n, i \neq j$. If S is a basis for the real inner product space V, then every vector v in V can be written as $v = p_1 v_1 + p_2 v_2 + \cdots p_n v_n$ for scalars $p_1, p_2, \ldots p_n$. Show that $p_i = \langle v, v_i \rangle$ for each i in the set $\{1, 2 \ldots. n\}$.

Solution

Consider the inner product $\langle v, v_i \rangle = \langle p_1 v_i + p_2 v_{2\ldots} p_n v_n, v \rangle$ for a certain i. Invoking the linearity property of inner product, we have

$$\langle v, v_i \rangle = p_1 \langle v_1, v_i \rangle + p_2 \langle v_2, v_i \rangle + \ldots. p_n \langle v_n, v_i \rangle.$$

Since $\langle v_i, v_j \rangle = 0$ for $i \neq j$ and i itself belongs to $\{1, 2, \ldots. n\}$, we have

$$\langle v, v_i \rangle = p_i \langle v_i, v_i \rangle$$

Again, $|v_i| = 1$, $\Rightarrow \langle v_i, v_i \rangle = 1$ because $|v_i| = \langle v_i, v_i \rangle^{1/2}$.

$$\Rightarrow \langle v, v_i \rangle = |v_i|^2 = 1$$

$$\Rightarrow \langle v, v_i \rangle = p_i.$$

This means that v can be written as

$$v = \langle v, v_1 \rangle v_1 + \langle v, v_2 \rangle v_2 + \cdots \langle v, v_n \rangle v_n.$$

This implies that we don't need to solve any system of linear equations in order to know the scalar coefficients. All we need do is calculate the inner products – provided that the basis is an orthonomal basis. Again,

if $S = \{v_1, v_2, \ldots v_n\}$ is an orthonormal basis for a inner product space V, then any vector v in V can be written as

$$v = \sum_{i=1}^{n} \langle v, v_i \rangle v_i, \text{ where } n \text{ is the dimension of the space.}$$

$$v = \sum_{i=1}^{n} \langle v, v_i \rangle v_i \qquad ----- P168$$

Example 2.24

Represent the vector $v = (2, -5)$ as a linear combination of the bases

$S_1 = \{(1,0), (0,1)\}$, and $S_2 = \left\{(\frac{1}{\sqrt{2}}, \frac{-1}{\sqrt{2}}), (\frac{1}{\sqrt{2}}, \frac{1}{\sqrt{2}})\right\}$ in R^2 with the standard inner product.

Solution

a)   It is not difficult to see that S_1 is an orthonormal basis. This is because

$$|(1,0)| = |(0,1)| = 1, and \ \langle(1,0), (0,1)\rangle = 0$$

Therefore,

$$v = \langle v, (0,1)\rangle(0,1) + \langle v, (0,1)\rangle(0,1)$$
$$\langle v, (0,1)\rangle = 2(0) + -5(1) = -5$$
$$\langle v, (1,0)\rangle = 2(1) + -5(0) = 2$$
$$\Rightarrow v = -5(0,1) + 2(1,0)$$

$$Or$$

$$v = 2(1,0) - 5(0,1)$$

Note that in general, for any v in V, the scalars p_1, p_2 such that
$v = p_1(1,0) + p_2(0,1)$ are not different from the component of v themselves;
that is, $p_1 = v_1$, and $p_2 = v_2$,where $v = (v_1, v_2)$. This is not unexpected
because any vector $v = (v_1, v_2)$ is initially relative to this special basis called the
standard basis. That is, any vector $v = (v_1, v_2)$ is naturally *coordinatized* with
respect to the standard basis S_1. Therefore, for an arbitrary basis
$S = \{v_1, v_2, ... v_n\}$, the *S-coordinatization* of any vector v in V is the vector
$v_s = (p_1, p_2, ... p_n)$, where $v = p_1 v_1 + p_2 v_2 + \cdots p_n v_n$. Therefore, writing a
vector v in V in terms of the basis vectors $v_1, v_2, ... v_n$ in S means we have
converted from the standard basis$\{(1,0,0 ... .0), (0, 1, 0, ...0), ... (0, 0, ...1)\}$ to
the basis $S = \{v_1, v_2, ... v_n\}$.
Again, if $v = p_1 v_1 + p_2 v_2 + \cdots p_n v_n$, then the *S-coordinatization* of v is

$$v_s = (p_1, p_2, ... p_n).$$

b) It is easily observed that S_2 is an orthonormal basis too. This is because

$$\left\langle \left(\frac{1}{\sqrt{2}}, \frac{-1}{\sqrt{2}}\right), \left(\frac{1}{\sqrt{2}}, \frac{1}{\sqrt{2}}\right)\right\rangle = 0 \text{ , and}$$

$$\left|\left(\frac{1}{\sqrt{2}}, \frac{-1}{\sqrt{2}}\right)\right| = \left|\left(\frac{1}{\sqrt{2}}, \frac{1}{\sqrt{2}}\right)\right| = 1$$

Therefore,

$$v = \left\langle v, \left(\frac{1}{\sqrt{2}}, \frac{-1}{\sqrt{2}}\right)\right\rangle \left(\frac{1}{\sqrt{2}}, \frac{-1}{\sqrt{2}}\right) + \left\langle v, \left(\frac{1}{\sqrt{2}}, \frac{1}{\sqrt{2}}\right)\right\rangle \left(\frac{1}{\sqrt{2}}, \frac{1}{\sqrt{2}}\right)$$

$$\left\langle v, \left(\frac{1}{\sqrt{2}}, \frac{-1}{\sqrt{2}}\right)\right\rangle = \frac{2}{\sqrt{2}} + \frac{5}{\sqrt{2}} = \frac{7\sqrt{2}}{2}$$

$$\left\langle v, \left(\frac{1}{\sqrt{2}}, \frac{1}{\sqrt{2}}\right)\right\rangle = \frac{2}{\sqrt{2}} - \frac{5}{\sqrt{2}} = \frac{-3\sqrt{2}}{2}$$

$$\Rightarrow \quad v = \frac{7\sqrt{2}}{2}\left(\frac{1}{\sqrt{2}}, \frac{-1}{\sqrt{2}}\right) + \frac{-3\sqrt{2}}{2}\left(\frac{1}{\sqrt{2}}, \frac{1}{\sqrt{2}}\right)$$

Therefore, the S_2- coordinatization of v is $v_{S_2} = \left(\frac{7\sqrt{2}}{2}, \frac{-3\sqrt{2}}{2}\right)$.

Now we give our attention to the systematic method of converting a given basis for an inner product space into an orthonormal basis for the same inner product space- **the Gram-Schmidt** process.

Essentially, the Gram-Schmidt process is a generalization of the concept of projecting a vector u along another vector v, and also orthogonal to v. Recall that the component of u parallel (along) v is

Projection of u on $v = \frac{\langle u, v \rangle}{\langle v, v \rangle} v =$ component of u along v or parallel to v.

Let the projection of u on $v = u_c \| v \ (or \ simply \ u \| v)$, and

Let the component of u orthogonal to $v = u_c \perp v$ (**or** simply $u \perp v$) , then

$$u = u_c \perp v + u_c \| v$$

$$u_c \| v = \frac{\langle u, v \rangle}{\langle v, v \rangle} v$$

$$\Rightarrow \quad u_c \perp v = u - \frac{\langle u,v \rangle}{\langle v,v \rangle} v = u \perp v$$

If $v = 1$, then $u_c \| v = \langle u, v \rangle \, v$, which is quite similar to what we have already established.

Moreover, $u_c \perp v$ is not just orthogonal to v only, but to every vector that is a linear multiple of v. In other words, $u_c \perp v$ is perpendicular to the linear span of v, a line. To generalize this concept, let us consider the following fact.

Let V be a real inner product space and let $S = \{v_1, v_2, \ldots v_n\}$ be a set of mutually orthogonal unit vectors from V. Let W be the subspace spanned by the set S; that is, W is the linear span of set $S(W = L(S))$. Then any vector u in V can be written in the form:

$$u = w_1 + w_2, \quad \text{where } w_1 \text{ is in } L(S) = W, \text{ and } w_2 \perp L(S).$$ We simply accomplish this by setting

$$w_1 = \langle w_1, v_1 \rangle \, v_1 + \langle w_1, v_2 \rangle v_2 + \ldots \ldots \langle w_1, v_k \rangle \, v_k$$
$$w_2 = u_c \perp L(S) = u_c - w_1.$$

In other words,

$$w_1 = \langle u - w_2, v_1 \rangle \, v_1 + \langle u - w_2, v_2 \rangle \, v_2 + \ldots \ldots \langle v - w_2, v_k \rangle \, v_k = u_c \| L(S)$$

Since $w_2 \perp L(S)$, and hence $w_2 \perp v_i$ for $1 \leq i \leq k$; therefore,

$$\langle w_2, v_i \rangle = 0 \text{ for each } v_i.$$

Thus,

$$u_c \| L(S) = \langle u, v_1 \rangle \, v_1 + \langle u, v_2 \rangle \, v_2 + \ldots \langle u, v_k \rangle v_k$$

This is the component of u in $L(S)$. Note that $L(S)$ is not a line in this case; thus, saying that '' the component of u parallel to $L(S)$ is not really appropriate.

In summary, u remains

$$u = u_c \perp L(S) + u_c \| L(S)$$

Where

$$u_c \| L(S) = \langle u, v_1 \rangle v_1 + \langle u, v_2 \rangle v_2 + \ldots \langle u, v_k \rangle v_k \qquad ---P169$$

When the set $S = \{v_1, v_2, \ldots v_n\}$ is a basis for V, it means that u itself is wholly in $L(S)$, and so $u_c \perp L(S) = 0$; it also means that $L(S) = $V. That is, if $L(S) = $V, then

$$\Rightarrow \quad u = u_c \| L(S) \quad , \quad u = \sum_{i=1}^{n} \langle u, v_i \rangle \, v_i$$

The above equation outlines the basic procedure used in the Gram-Schmidt process. Suppose that $\{v_1, v_2\}$ is a basis of an inner product space V, then $v_i \neq 0$, and hence $|v_1| \neq 0$. Let $w_1 = \frac{v_1}{|v_1|}$, then w_1 is a unit vector basis for a one-dimensional subspace W of V. Now let w_2' be the component of $v_2 \perp w_1$.

$$w_2' = v_2 - \langle v_2, w_1 \rangle \, w_1.$$

Let

$$w_2 = \frac{w_2'}{|w_2'|} = \frac{v_2 - \langle v_2, w_1 \rangle}{|v_2 - \langle v_2, w_1 \rangle|}$$

Therefore,

$$|w_2| = 1.$$

The set $\{w_1, w_2\}$ forms an orthonormal basis for V. This is the Gram-Schmidt process for a two-dimensional inner product space V.

Similarly, for a three-dimensional space, let the set $S = \{v_1, v_2, v_3\}$ be a basis for an inner product space, then

$$w_1 = \frac{v_1}{|v_1|}$$

$$w_2' = v_2 - \langle v_2, w_1 \rangle w_1$$

$$w_2 = \frac{w_2'}{|w_2'|} = \frac{v_2 - \langle v_2, w_1 \rangle \, w_1}{|v_2 - \langle v_2, w_1 \rangle \, w_1|} \qquad\qquad ----- P169$$

$$w_3' = v_3 - \langle v_3, w_1 \rangle \, w_1 - \langle v_3, w_2 \rangle \, w_2$$

$$w_3 = \frac{w_3'}{|w_3'|} = \frac{v_3 - \langle v_3, w_1 \rangle\, w_1 - \langle v_3, w_2 \rangle\, w_2}{|\, v_3 - \langle v_3, w_1 \rangle\, w_1 - \langle v_3, w_2 \rangle\, w_2|} \quad ----- P170$$

The set $\{w_1, w_2, w_3\}$ is an orthonormal basis for V. This is the Gram-Schmidt process for 3-dimensional inner product space V. To illustrate it, let us consider the following examples.

Example2.25

Transform the basis $S = \{(1,1), (1,-1)\}$ for V into an orthonormal basis for V with respect to the Euclidean inner product.

Solution

We follow the Gram-Schmidt procedure by first normalizing the vector $(1,1)$.

$$w_1 = \frac{(1,1)}{|(1,1)|} = \frac{(1,1)}{\sqrt{2}}$$

$$w_2' = (1,-1) - \langle (1,-1), \frac{(1,1)}{\sqrt{2}} \rangle \frac{(1,1)}{\sqrt{2}}$$

$$w_2' = (1,-1) - (0,0)$$

$$= (1,-1)$$

$$w_2 = \frac{w_2'}{|w_2'|} = \frac{(1,-1)}{|(1,-1)|}$$

$$w_2 = \frac{(1,-1)}{\sqrt{2}}$$

Therefore, the set $S' = \left\{ \frac{(1,1)}{\sqrt{2}}, \frac{(1,-1)}{\sqrt{2}} \right\}$ is an orthonormal basis for V.

Example 2.26

Transform the basis $S = \{(3,2), (4,-1)\}$ in to an Orthonormal basis for V with respect to the Euclidean inner product.

Solution

As before,

$$w_1 = \frac{(3,2)}{|(3,2)|} = \frac{(3,2)}{\sqrt{13}}$$

$$w_2' = (4,-1) - \langle (4,-1), \frac{(3,2)}{\sqrt{13}} \rangle \frac{(3,2)}{\sqrt{13}}$$

$$w_2' = (4,-1) - \frac{10}{13}(3,2)$$

$$w_2' = \left(\frac{22}{13}, \frac{-33}{13}\right)$$

$$w_2 = \frac{w_2'}{w_2'} = \frac{(22,-33)}{|(22,-33)|}$$

$$w_2 = \frac{(22,-33)}{\sqrt{1573}}$$

Therefore, the set $s' = \left\{ \frac{(3,2)}{\sqrt{13}}, \frac{(22,-33)}{\sqrt{1573}} \right\}$ is an orthonormal basis for V.

Example 2.27

Transform the basis $S = \{(1,2,0), (-1,0,3), (1,1,1)\}$ for an inner product space V into an orthonormal basis s' for V with respect to the Euclidean inner product.

Solution

$$w_1 = \frac{(1,\ 2,\ 0)}{|(1,\ 2,\ 0)|} = \frac{(1,\ 2,\ 0)}{\sqrt{5}}$$

$$w_2' = (-1,0,3) - \langle (-1,0,3), \frac{(1,2,0)}{\sqrt{5}} \rangle \frac{(1,2,0)}{\sqrt{5}}$$

$$= (-1,0,3) + \frac{(1,\ 2,0)}{5}$$

$$= \frac{1}{5}(-4,2,15)$$

$$w_2 = \frac{w_2'}{w_2'} = \frac{\frac{1}{5}(-4,2,15)}{\left|\frac{1}{5}(-4,2,15)\right|} = \frac{(-4,2,15)}{\sqrt{245}}$$

$$w_3' = (1,1,1) - \left\langle (1,1,1), \frac{(-4,2,15)}{\sqrt{245}} \right\rangle \frac{(-4,2,15)}{\sqrt{245}} - \left\langle (1,1,1), \frac{(1,2,0)}{\sqrt{5}} \right\rangle \frac{(1,2,0)}{\sqrt{5}}$$

$$w_3' = (1,1,1) - \frac{13}{245}(-4,2,15) - \frac{1}{5}(1,2,0)$$

$$w_3' = \left(\frac{30}{49}, \frac{-15}{49}, \frac{10}{49} \right)$$

$$w_3 = \frac{w_3'}{w_3'} = \frac{(30, -15, 10)}{|(30, -15, 10)|} = \frac{(30, -15, 10)}{\sqrt{1225}}$$

$$w_3 = \frac{(30, -15, 10)}{35}$$

$$w_3 = \frac{(6,-3,2)}{7}$$

An orthonormal set S' for V is $S' = \left\{ \frac{(1,2,0)}{\sqrt{5}}, \frac{(-4,2,15)}{\sqrt[7]{5}}, \frac{(6,-3,2)}{7} \right\}$.

Example 2.28

Find an orthonormal basis S' for the inner product space V spanned by the basis $S = \{(1,0,0,1), (0,-1,0,1), (0,0,1,1)\}$ with respect to the Euclidean inner product.

Solution

$$w_1 = \frac{(1,0,0,1)}{|(1,0,01)|} = \frac{(1,0,0,1)}{\sqrt{2}}$$

$$w_2' = (0,-1,0,1) - \left\langle (0,-1,0,1), \frac{(1,0,0,1)}{\sqrt{2}} \right\rangle \frac{(1,0,0,1)}{\sqrt{2}}$$

$$w_2' = (0,-1,0,1) - \frac{(1,0,0,1)}{2}$$

$$= \left(\frac{-1}{2}, -1, 0, \frac{1}{2} \right) = \frac{-1}{2}(1,2,0,-1)$$

$$w_2 = \frac{w_2'}{|w_2'|} = \frac{(1,2,0,-1)}{|(1,2,0,-1)|}$$

$$w_2 = \frac{(1,2,0,-1)}{\sqrt{6}}$$

$$w_3' = (0,0,1,1) - \left\langle (0,0,1,1), \frac{(1,2,0,-1)}{\sqrt{6}} \right\rangle \frac{(1,2,0,-1)}{\sqrt{6}} - \left\langle (0,0,1,1), \frac{(1,0,01)}{\sqrt{2}} \right\rangle \frac{(1,0,01)}{\sqrt{2}}$$

$$w_3' = (0,0,1,1) + \frac{(1,2,0,-1)}{6} - \frac{(1,0,01)}{2}$$

$$= \left(\frac{-1}{3}, \frac{1}{3}, 1, \frac{1}{3} \right)$$

$$= \frac{1}{3}(-1,1,3,1)$$

$$w_3 = \frac{w_3'}{|w_3'|} = \frac{(-1,1,3,1)}{|(-1,1,3,1)|}$$

$$w_3 = \frac{(-1,1,3,1)}{2\sqrt{3}}$$

Therefore, the set $S' = \left\{ \frac{(1,0,0,1)}{\sqrt{2}}, \frac{(1,2,0,-1)}{\sqrt{6}}, \frac{(-1,1,3,1)}{2\sqrt{3}} \right\}$ is an orthonormal basis for V.

In general, if $S = \{v_1, v_2, \dots v_n\}$ is a basis for V, and suppose that we have already constructed, step-by-step, orthonormal vectors $w_1, w_2, \dots w_{n-1}$ having the same linear span as $v_1, v_2, \dots v_{n-1}$, the subspace W. Then the component of v_n orthogonal to W is

$$v_{n,c} \perp w = v_n - \sum_{i=1}^{n-1} \langle v_n, w_i \rangle w_i = w_n'$$

$$\therefore w_n = \frac{w_n'}{|w_n'|}$$

$$\Rightarrow \qquad w_n = \frac{v_n - \sum_{i=1}^{n-1} \langle v_n, w_i \rangle w_i}{\left| v_n - \sum_{i=1}^{n-1} \langle v_n, w_i \rangle w_i \right|} \qquad \qquad ----- P171$$

Note that $w_n \neq 0$. For if $w_n = 0$, then according to $P171$

$$v_n = \sum_{i=1}^{n-1} \langle v_n, w_i \rangle w_i .$$ This implies that v_n is in the linear span of

$w_1, w_2 \ldots w_{n-1}$, and hence the linear span of $v_1, v_2 \ldots v_{n-1}$. This is contrary to the initial assumption that $S = \{v_1, v_2 \ldots v_n\}$ is a basis for V. Therefore, $w_n \neq 0$.

Example 2.29

Convert basis $S = \{(1,0,1,0), (0,1,0,1), (1,0,-1,0), (1,-1,-1,1)\}$ for R^n into an orthonormal basis with respect to the standard (Euclidean) inner product in R^n.

Solution

$$w_1 = \frac{(1,0,1,0)}{|(1,0,1,0)|} = \frac{(1,0,1,0)}{\sqrt{2}}$$

$$w_2' = (0,1,0,1) - \langle(0,1,0,1), \frac{(1,0,1,0)}{\sqrt{2}}\rangle\frac{(1,0,1,0)}{\sqrt{2}}$$

$$= (0,1,0,1) - (0,0,0,0)$$

$$= (0,1,0,1)$$

$$w_2 = \frac{w_2'}{|w_2'|} = \frac{(0,1,0,1)}{\sqrt{2}}$$

$$w_3' = (1,0,-1,0) - \langle(1,0,-1,0), \frac{(0,1,0,1)}{\sqrt{2}}\rangle\frac{(0,1,0,1)}{\sqrt{2}} - \langle(1,0,-1,0), \frac{(1,0,1,0)}{\sqrt{2}}\rangle\frac{(1,0,1,0)}{\sqrt{2}}$$

$$= (1,0,-1,0) - (0,0,0,0) - (0,0,0,0)$$

$$= (1,0,-1,0)$$

$$w_3 = \frac{(1,0,-1,0)}{\sqrt{2}}$$

$$w_4' = (1,-1,-1,1) - \langle(1,-1,-1,1), \frac{(1,0,-1,0)}{\sqrt{2}}\rangle\frac{(1,0,-1,0)}{\sqrt{2}} - \langle(1,-1,-1,1), \frac{(0,1,0,1)}{\sqrt{2}}\rangle\frac{(0,1,0,1)}{\sqrt{2}}$$

$$- \langle(1,-1,-1,1), \frac{(1,0,1,0)}{\sqrt{2}}\rangle\frac{(1,0,1,0)}{\sqrt{2}}$$

$$w_4' = (1,-1,-1,1) - (1,0,-1,0) - (0,0,0,0) - (0,0,0,0)$$

$$= (0,-1,0,1)$$

$$w_4 = \frac{(0,-1,0,1)}{|(0,-1,0,1)|} = \frac{(0,-1,0,1)}{\sqrt{2}}$$

An orthonormal basis for R^4 is $S' = \left\{\frac{(1,0,1,0)}{\sqrt{2}}, \frac{(0,1,0,1)}{\sqrt{2}}, \frac{(1,0,-1,0)}{\sqrt{2}}, \frac{(0,-1,0,1)}{\sqrt{2}}\right\}$.

Example 2.30

Convert the basis $S = \{3, 2x - 1, x^2 + 1\}$ for P_2 into an orthonormal basis under the integral inner product.

Solution

In this case, $v_1 = 3$

w_1, as before, is given by

$$w_1 = \frac{v_1}{|v_1|}$$

$$|v_1| = \langle v_1, v_1 \rangle^{\frac{1}{2}} = \left(\int_0^1 3.3 \, dx \right)^{\frac{1}{2}}$$

$$\left(\int_0^1 9 \, dx \right)^{\frac{1}{2}} = (9)^{\frac{1}{2}} = 3$$

$\Rightarrow \quad w_1 = \frac{3}{3} = 1$; this is always the case for any polynomial of zero degree.

$$w_2' = v_2 - \langle v_2, w_1 \rangle w_1$$

$$\langle v_2, w_1 \rangle = \int_0^1 (2x - 1).1 \, dx$$

$$= |x^2 - x|_0^1 = 0$$

$$\Rightarrow \quad w_2' = v_2$$

$$w_2 = \frac{w_2'}{|w_2'|} = \frac{v_2}{|v_2|}$$

But

$$|v_2| = \langle v_2, v_2 \rangle^{\frac{1}{2}} = \left(\int_0^1 (2x - 1)^2 dx \right)^{\frac{1}{2}}$$

$$= \left(\frac{1}{2} | \frac{(2x - 1)^3}{3} |_0^1 \right)^{\frac{1}{2}}$$

$$|v_2| = \frac{1}{\sqrt{3}}$$

$$w_2 = \frac{2x-1}{\frac{1}{\sqrt{3}}} = \sqrt{3}\,(2x-1)$$

$$w_3' = v_3 - \langle v_3, \sqrt{3}(2x-1)\rangle \sqrt{3}\,(2x-1) - \langle v_3, 1\rangle (1)$$

$$\langle v_3, \sqrt{3}(2x-1)\rangle = \int_0^1 (x^2+1)\sqrt{3}\,(2x-1)dx$$

$$= \sqrt{3}\left|\frac{x^4}{2} - \frac{x^3}{3} + \frac{x^2}{2} - x\right|_0^1$$

$$= \frac{-1}{\sqrt{3}}$$

$$\langle v_3, 1\rangle = \int_0^1 (x^2+1)dx$$

$$\left|\frac{x^3}{3} + x\right|_0^1 = \frac{4}{3}$$

$$\Rightarrow \quad w_3' = (x^2+1) - \left(\frac{-1}{\sqrt{3}}\right)\sqrt{3}\,(2x-1) - \frac{4}{3}$$

$$= x^2 + 2x - \frac{4}{3}$$

$$w_3 = \frac{w_3'}{|w_3'|} = \frac{x^2 + 2x - \frac{4}{3}}{\left(\int_0^1 (x^2 + 2x - \frac{4}{3})^2 dx\right)^{\frac{1}{2}}}$$

$$w_3 = \frac{x^2 + 2x - \frac{4}{3}}{\left(\int_0^1 \left(x^4 + 4x^3 + \frac{4x^2}{3} - \frac{16x}{3} + \frac{16}{9}\right)dx\right)^{\frac{1}{2}}}$$

$$w_3 = \sqrt{\frac{45}{34}}\left(x^2 + 2x - \frac{4}{3}\right)$$

The set $S' = \{1, \sqrt{3}(2x-1), \sqrt{\frac{45}{34}}\,(x^2 + 2x - \frac{4}{3})\}$ is an orthonormal basis for P_2 under the integral inner product.

One more significance of orthonormalizing a given basis for a given inner product space is illustrated in the following fact.

Let V be any real inner product space and let $S = \{v_1, v_2 \dots v_n\}$ be an orthonormal basis for V. If

$u = a_1 v_1 + a_2 v_2 + \dots\dots a_n v_n$ *and* $w = b_1 v_1 + b_2 v_2 + \dots\dots b_n v_n$ *are any vectors in V, then*

$$\langle u, w \rangle = a_1 b_1 + a_2 b_2 + \cdots a_n b_n.$$

Proof

$$\langle u, w \rangle = \langle a_1 v_1 + a_2 v_2 + \cdots a_n v_n \;,\; b_1 v_1 + b_2 v_2 + \dots b_n v_n \rangle.$$

Invoking the linearity property of the inner product, we have

$$\langle u, w \rangle = \sum_{j=1}^{n} \sum_{i=1}^{n} a_i b_j \langle v_i, v_j \rangle$$

Since

$$\langle v_i, v_j \rangle = 0 \;,\; for\, i \neq j$$

$$\langle u, w \rangle = \sum_{i=1}^{n} a_i b_i \langle v_i, v_i \rangle$$

Again

$$\langle v_i, v_i \rangle = 1 \text{ for each } i.$$

$$\Rightarrow \langle u, w \rangle = \sum_{i=1}^{n} a_i b_i$$

APPLICATION TO LEAST SQUARE APPROXIMATION

There is often the need to plot the data result of an experiment. Due to the imperfect nature of the experimenter, some errors might emerge, which may lead to a scatter diagram.

By scatter diagram we mean points in a plane that are not collinear (in the case of a bivariate data). If we expect the graph representing the experimental data to be a straight line, then we are challenged to **fit** in a straight line that best represents the data. Keep in mind that this line of best fit may not even pass through any of the data points. The line of best fit is also called the **least square** line because it minimizes the sum of the squares of the vertical errors as will be seen shortly.

Suppose that the n data points $(x_1, y_1), (x_2, y_2), \dots (x_n, y_n)$ are experimentally determined. Through these points, we wish to find a straight line:

$$y = ax + b$$

If these points do indeed fall on the line, then we have the following equations in the two unknowns a and b.

$$y_1 = ax_1 + b$$
$$y_2 = ax_2 + b$$
$$\vdots \qquad \vdots \qquad \vdots$$
$$y_n = ax_n + b$$

which, in matrix representation, is

$$\begin{pmatrix} y_1 \\ y_2 \\ \cdot \\ \cdot \\ \cdot \\ y_n \end{pmatrix} = \begin{pmatrix} x_1 & 1 \\ x_2 & 1 \\ \cdot & \cdot \\ \cdot & \cdot \\ x_n & 1 \end{pmatrix} \begin{pmatrix} a \\ b \end{pmatrix}$$

Or in abbreviated matrix form

$$Y = MS$$

Where

$$Y = \begin{pmatrix} y_1 \\ y_2 \\ \cdot \\ \cdot \\ \cdot \\ y_n \end{pmatrix} , \; S = \begin{pmatrix} a \\ \cdot \\ \cdot \\ \cdot \\ b \end{pmatrix} , \; M = \begin{pmatrix} x_1 & 1 \\ x_2 & 1 \\ \cdot & \cdot \\ \cdot & \cdot \\ x_n & 1 \end{pmatrix}$$

This system is expected to be inconsistent because the data points are not collinear for they are experimentally determined. Therefore, we seek a unique vector $S' = \begin{pmatrix} a' \\ b' \end{pmatrix}$, such that $|Y - MS'|$ is minimum ; that is, we seek the straight line $y = a'x + b'$. This is the line of best fit.

Note that minimizing $|Y - MS'|$ is the same as minimizing $|Y - MS'|^2$, and that

$$|Y - MS'|^2 = (y_1 - a'x_1 - b')^2 + (y_2 - a'x_2 - b')^2 + \cdots (y_n - a'x_n - b')^2,$$

where $d_i = |y_i - a'x_i - b'| = $ vertical distance between $(x_i, y_i) \; and \, (x_i', y_i')$, where $y_i' = a'x_i + b'$.

$$|Y - MS'|^2 = \sum_{i=1}^{n} d_i^2 \, , \qquad \text{hence the name least square.}$$

To calculate S' that minimizes $|Y - MS|$, we first note that Y is a fixed vector in $\mathbb{R}^n$, and S varies over all possible values, making the vector MS to also vary over all possible values, and thus, forming a subspace of $\mathbb{R}^n$, the column space W of the matrix M.

$$MS = a \begin{pmatrix} x_1 \\ x_2 \\ \cdot \\ \cdot \\ \cdot \\ x_n \end{pmatrix} + b \begin{pmatrix} 1 \\ 1 \\ \cdot \\ \cdot \\ \cdot \\ 1 \end{pmatrix}.$$

Therefore, for the vector $Y - MS'$ to be minimum, it must be orthogonal to the column space of M, the linear span of the set

$$S = \{(x_1, x_2, \dots x_n), (1, 1, \dots 1)\}.$$

This means that

$$\langle Y - MS', MS \rangle = 0$$

$$\Rightarrow \quad (MS)^T (Y - MS') = 0$$

$$\Rightarrow \quad S^T M^T (Y - MS') = 0$$

$$\Rightarrow \quad S^T (M^T Y - M^T MS') = 0$$

$$\Rightarrow \quad \langle M^T Y - M^T MS', S \rangle = 0$$

Since $S \neq 0$, therefore,

$$M^T Y - M^T MS' = 0$$

$$M^T MS' = M^T Y$$

$$S' = (M^T M)^{-1} M^T Y \qquad\qquad ----- P172$$

Example 2.31

Find the straight line that is the least – square fit to the points $(0,1), (-1, 2)(1,4), (3,5)$.

Solution

Let the line of best fit be

$$y = a'x + b'$$

The given data points correspond to the system of linear equations in a and b as follows

$$1 = b$$

$$2 = -a + b$$

$$4 = a + b$$

$$5 = 3a + b$$

$$\Rightarrow \quad \begin{pmatrix} 0 & 1 \\ -1 & 1 \\ 1 & 1 \\ 3 & 1 \end{pmatrix} \begin{pmatrix} a \\ b \end{pmatrix} = \begin{pmatrix} 1 \\ 2 \\ 4 \\ 5 \end{pmatrix}$$

$$MS = Y$$

If $S' = \begin{pmatrix} a' \\ b' \end{pmatrix}$ is the minimizer of $|Y - MS|$, then

$$S' = (M^T M)^{-1} M^T Y$$

$$M^T M = \begin{pmatrix} 0 & -1 & 1 & 3 \\ 1 & 1 & 1 & 1 \end{pmatrix} \begin{pmatrix} 0 & 1 \\ -1 & 1 \\ 1 & 1 \\ 3 & 1 \end{pmatrix}$$

$$= \begin{pmatrix} 0+1+1+9 & 0-1+1+3 \\ 0-1+1+3 & 1+1+1+1 \end{pmatrix} = \begin{pmatrix} 11 & 3 \\ 3 & 4 \end{pmatrix}$$

$$(M^T M)^{-1} = \frac{1}{35} \begin{pmatrix} 4 & -3 \\ -3 & 11 \end{pmatrix}$$

$$(M^T M)^{-1} M^T = \frac{1}{35} \begin{pmatrix} 4 & -3 \\ -3 & 11 \end{pmatrix} \begin{pmatrix} 0 & -1 & 1 & 3 \\ 1 & 1 & 1 & 1 \end{pmatrix}$$

$$= \frac{1}{35} \begin{pmatrix} -3 & -7 & 1 & 9 \\ 11 & 14 & 8 & 2 \end{pmatrix}$$

$$S' = \frac{1}{35} \begin{pmatrix} -3 & -7 & 1 & 9 \\ 11 & 14 & 8 & 2 \end{pmatrix} \begin{pmatrix} 1 \\ 2 \\ 4 \\ 5 \end{pmatrix}$$

$$S' = \frac{1}{35} \begin{pmatrix} -3-14+4+45 \\ 11+28+32+10 \end{pmatrix}$$

$$S' = \frac{1}{35} \begin{pmatrix} 32 \\ 81 \end{pmatrix} = \begin{pmatrix} a' \\ b' \end{pmatrix}$$

$$\Rightarrow \quad a' = \frac{32}{35}, \; b' = \frac{81}{35}$$

The line of best fit is $y = \frac{32}{35} x + \frac{81}{35}$

Note that we can generalize the above method to an equation of the form

$$y = a_1 x_1 + a_2 x_2 + \cdots a_n x_n$$

If the n data points are substituted, we have

$$y_1 = a_1 x_{11} + a_2 x_{21} + \cdots a_n x_{n1}$$
$$y_2 = a_1 x_{12} + a_2 x_{22} + \cdots a_n x_{n2}$$

$$y_n = a_1 x_{1n} + a_2 x_{2n} + \cdots a_n x_{nn}$$

which can be written in matrix form as

$$\begin{pmatrix} y_1 \\ y_2 \\ \vdots \\ y_n \end{pmatrix} = \begin{pmatrix} x_{11} & x_{21} & \cdots\cdots x_{n1} \\ x_{12} & x_{22} & \cdots\cdots x_{n2} \\ x_{1n} & x_{2n} & \cdots\cdots x_{nn} \end{pmatrix} \begin{pmatrix} a_1 \\ a_2 \\ \\ a_n \end{pmatrix}$$

$$Y = MS$$

We can use the same argument to show that the minimizing vector S' is given by

$$S' = \begin{pmatrix} a'_1 \\ a'_2 \\ \vdots \\ \vdots \\ a'_n \end{pmatrix} = (M^T M)^{-1} M^T Y$$

Again, if we expect a polynomial curve of degree k, other than a straight line, we have

$$y = a_n x^k + a_{n-1} x^{k-1} + \cdots a_0, \qquad k > 1$$

when we substitute the data points, as usual, we have

$$y = a_n x_1^k + a_{n-1} x_1^{k-1} + \cdots a_0$$
$$y = a_n x_2^k + a_{n-1} x_2^{k-1} + \cdots a_0$$

$$y = a_n x_n^k + a_{n-1} x_n^{k-1} + \cdots a_0$$

In matrix form, this gives

$$\begin{pmatrix} y_1 \\ y_2 \\ \vdots \\ y_n \end{pmatrix} = \begin{pmatrix} x_1^k & x_1^{k-1} & \cdots & 1 \\ x_2^k & x_2^{k-1} & \cdots & 1 \\ x_n^k & x_n^{k-1} & \cdots & 1 \end{pmatrix} \begin{pmatrix} a_n \\ a_{n-1} \\ \vdots \\ a_0 \end{pmatrix}$$

$$Y = MS$$

$$S' = \begin{pmatrix} a_n' \\ a_{n-1}' \\ \vdots \\ \vdots \\ a_0' \end{pmatrix} = (M^T M)^{-1} M^T Y$$

Example 2.32

If the variables $x_1, x_2, and\ y$ are known to be linearly related as

$y = a_1 x_1 + a_2 x_2$, then obtain the line of best fit for the following experimental data.

y	x_1	x_2
2	1	3
0	3	2
3	-1	4
1	2	1
4	5	7

Solution

$$y = a_1 x_1 + a_2 x_2$$
$$\Rightarrow \quad 2 = a_1 + 3a_2$$
$$0 = 3a_1 + 2a_2$$
$$3 = -a_1 + 4a_2$$
$$1 = 2a_1 + a_2$$
$$4 = 5a_1 + 7a_2$$

$$\begin{pmatrix} 2 \\ 0 \\ 3 \\ 1 \\ 4 \end{pmatrix} = \begin{pmatrix} 1 & 3 \\ 3 & 2 \\ -1 & 4 \\ 2 & 1 \\ 5 & 7 \end{pmatrix} \begin{pmatrix} a_1 \\ a_2 \end{pmatrix}$$

$$Y = MS$$

If $S' = \begin{pmatrix} a_1' \\ a_2' \end{pmatrix}$, the minimizer, then

$$S' = (M^T M)^{-1} M^T Y$$

$$M^T M = \begin{pmatrix} 1 & 3 & -1 & 2 & 5 \\ 3 & 2 & 4 & 1 & 7 \end{pmatrix} \begin{pmatrix} 1 & 3 \\ 3 & 2 \\ -1 & 4 \\ 2 & 1 \\ 5 & 7 \end{pmatrix}$$

$$= \begin{pmatrix} 1+9+1+4+25 & 3+6-4+2+35 \\ 3+6-4+2+35 & 9+4+16+1+49 \end{pmatrix}$$

$$M^T M = \begin{pmatrix} 40 & 42 \\ 42 & 79 \end{pmatrix}$$

$$(M^T M)^{-1} = \frac{1}{1396} \begin{pmatrix} 79 & -42 \\ -42 & 40 \end{pmatrix}$$

$$(M^T M)^{-1} M^T = \frac{1}{1396} \begin{pmatrix} 79 & -42 \\ -42 & 40 \end{pmatrix} \begin{pmatrix} 1 & 3 & -1 & 2 & 5 \\ 3 & 2 & 4 & 1 & 7 \end{pmatrix}$$

$$= \frac{1}{1396} \begin{pmatrix} -47 & 153 & -247 & 116 & 101 \\ 78 & -48 & 202 & -44 & 70 \end{pmatrix}$$

$$S' = \frac{1}{1396} \begin{pmatrix} -47 & 153 & -247 & 116 & 101 \\ 78 & -48 & 202 & -44 & 70 \end{pmatrix} \begin{pmatrix} 2 \\ 0 \\ 3 \\ 1 \\ 4 \end{pmatrix}$$

$$S' = \frac{1}{1396} \begin{pmatrix} -94+0-741+116+404 \\ 156+0+606-44+280 \end{pmatrix}$$

$$S' = \frac{1}{1396}\begin{pmatrix} -315 \\ 998 \end{pmatrix}$$

Therefore, the line of best fit is

$$y = \frac{-315}{1396}x_1 + \frac{998}{1396}x_2$$

Example 2.33

Find the quadratic equation $y = a_2x^2 + a_1x + a_0$ that is the least square fit to the following set of data points.

$(-2,1), (-1,3), (0,5), (1,4)$.

Solution

$$4a_2 - 2a_1 + a_0 = 1$$
$$a_2 - a_1 + a_0 = 3$$
$$a_0 = 5$$
$$a_2 + a_1 + a_0 = 4$$

$$\begin{pmatrix} 4 & -2 & 1 \\ 1 & -1 & 1 \\ 0 & 0 & 1 \\ 1 & 1 & 1 \end{pmatrix}\begin{pmatrix} a_2 \\ a_1 \\ a_0 \end{pmatrix} = \begin{pmatrix} 1 \\ 3 \\ 5 \\ 4 \end{pmatrix}$$

$$MS = Y$$

As before,

$$S' = \begin{pmatrix} a_2' \\ a_1' \\ a_0' \end{pmatrix} = (M^TM)^{-1}M^TY$$

$$M^TM = \begin{pmatrix} 4 & 1 & 0 & 1 \\ -2 & -1 & 0 & 1 \\ 1 & 1 & 1 & 1 \end{pmatrix}\begin{pmatrix} 4 & -2 & 1 \\ 1 & -1 & 1 \\ 0 & 0 & 1 \\ 1 & 1 & 1 \end{pmatrix}$$

$$= \begin{pmatrix} 18 & -8 & 6 \\ -8 & 6 & -2 \\ 6 & -2 & 4 \end{pmatrix}$$

$$(M^T M)^{-1} = \begin{pmatrix} \dfrac{1}{4} & \dfrac{1}{4} & -\dfrac{1}{4} \\[2mm] \dfrac{1}{4} & \dfrac{9}{20} & \dfrac{-3}{20} \\[2mm] -\dfrac{1}{4} & \dfrac{-3}{20} & \dfrac{11}{20} \end{pmatrix}$$

$$(M^T M)^{-1} M^T = \begin{pmatrix} \dfrac{1}{4} & \dfrac{1}{4} & -\dfrac{1}{4} \\[2mm] \dfrac{1}{4} & \dfrac{9}{20} & \dfrac{-3}{20} \\[2mm] -\dfrac{1}{4} & \dfrac{-3}{20} & \dfrac{11}{20} \end{pmatrix} \begin{pmatrix} 4 & 1 & 0 & 1 \\ -2 & -1 & 0 & 1 \\ 1 & 1 & 1 & 1 \end{pmatrix}$$

$$(M^T M)^{-1} M^T = \begin{pmatrix} \dfrac{1}{4} & \dfrac{-1}{4} & \dfrac{-1}{4} & \dfrac{1}{4} \\[2mm] \dfrac{-1}{20} & \dfrac{-7}{20} & \dfrac{-3}{20} & \dfrac{11}{20} \\[2mm] \dfrac{-3}{20} & \dfrac{9}{20} & \dfrac{11}{20} & \dfrac{3}{20} \end{pmatrix}$$

Finally,

$$S' = (M^T M)^{-1} M^T Y = \begin{pmatrix} \dfrac{1}{4} & \dfrac{-1}{4} & \dfrac{-1}{4} & \dfrac{1}{4} \\[2mm] \dfrac{-1}{20} & \dfrac{-7}{20} & \dfrac{-3}{20} & \dfrac{11}{20} \\[2mm] \dfrac{-3}{20} & \dfrac{9}{20} & \dfrac{11}{20} & \dfrac{3}{20} \end{pmatrix} \begin{pmatrix} 1 \\ 3 \\ 5 \\ 4 \end{pmatrix}$$

$$S' = \begin{pmatrix} -3/4 \\ 7/20 \\ 91/20 \end{pmatrix}$$

$$\Rightarrow \quad a'_2 = \dfrac{-3}{4}, \; a'_1 = \dfrac{7}{20}, \; a'_0 = \dfrac{91}{20}$$

Therefore, the quadratic curve of best fit is

$$y = \frac{-3}{4}x^2 + \frac{7}{20}x + \frac{91}{20}$$

Example 2.34

Find the cubic polynomial $y = a_3x^3 + a_2x^2 + a_1x + a_0$ that is the least square fit to the data points: $(-2,-8),(-1,-1),(0,3),(1,1),(2,-1),(3,0)$

Solution

As before,

$$M = \begin{pmatrix} 8 & 4 & -2 & 1 \\ 1 & 1 & -1 & 1 \\ 0 & 0 & 0 & 1 \\ 1 & 1 & 1 & 1 \\ 8 & 4 & 2 & 1 \\ 27 & 9 & 3 & 1 \end{pmatrix} \begin{pmatrix} a_3 \\ a_2 \\ a_1 \\ a_0 \end{pmatrix} = \begin{pmatrix} -2 \\ -1 \\ 0 \\ 1 \\ 2 \\ 3 \end{pmatrix}$$

$$S' = (M^TM)^{-1}M^TY$$

$$M^TM = \begin{pmatrix} 8 & 1 & 0 & 1 & 8 & 27 \\ 4 & 1 & 0 & 1 & 4 & 9 \\ -2 & -1 & 0 & 1 & 2 & 3 \\ 1 & 1 & 1 & 1 & 1 & 1 \end{pmatrix} \begin{pmatrix} 8 & 4 & -2 & 1 \\ 1 & 1 & -1 & 1 \\ 0 & 0 & 0 & 1 \\ 1 & 1 & 1 & 1 \\ 8 & 4 & 2 & 1 \\ 27 & 9 & 3 & 1 \end{pmatrix}$$

$$= \begin{pmatrix} 859 & 309 & 81 & 45 \\ 309 & 115 & 27 & 19 \\ 81 & 27 & 19 & 3 \\ 45 & 19 & 3 & 6 \end{pmatrix}$$

$$(M^T M)^{-1} = \begin{pmatrix} \dfrac{35}{473} & \dfrac{-405}{1892} & \dfrac{-63}{1892} & \dfrac{6}{43} \\[2ex] \dfrac{-405}{1892} & \dfrac{553}{856} & \dfrac{921}{13244} & \dfrac{-143}{301} \\[2ex] \dfrac{-63}{1892} & \dfrac{921}{13244} & \dfrac{282}{2851} & \dfrac{-6}{301} \\[2ex] \dfrac{6}{43} & \dfrac{-143}{301} & \dfrac{-6}{301} & \dfrac{191}{301} \end{pmatrix}$$

$$(M^T M)^{-1} M^T = \begin{pmatrix} \dfrac{-5}{86} & \dfrac{31}{946} & \dfrac{6}{43} & \dfrac{-16}{473} & \dfrac{-181}{946} & \dfrac{105}{946} \\[2ex] \dfrac{155}{602} & \dfrac{-373}{3311} & \dfrac{-143}{301} & \dfrac{25}{946} & \dfrac{759}{1417} & \dfrac{-207}{893} \\[2ex] \dfrac{-62}{301} & \dfrac{-547}{6622} & \dfrac{-6}{301} & \dfrac{109}{946} & \dfrac{628}{3311} & \dfrac{12}{3311} \\[2ex] \dfrac{-33}{301} & \dfrac{96}{301} & \dfrac{191}{301} & \dfrac{12}{43} & \dfrac{-57}{301} & \dfrac{20}{301} \end{pmatrix}$$

Finally,

$$S' = (M^T M)^{-1} M^T Y$$

$$S' = (M^T M)^{-1} M^T Y \begin{pmatrix} -2 \\ -1 \\ 0 \\ 1 \\ 2 \\ 3 \end{pmatrix}$$

$$S' = \begin{pmatrix} \dfrac{1}{6004799503160661} \\[2ex] \dfrac{1}{643371375338642} \\[2ex] 1 \\[2ex] \dfrac{1}{3003299751580331} \end{pmatrix} = \begin{pmatrix} a'_3 \\ a'_2 \\ a'_1 \\ a'_0 \end{pmatrix}$$

Therefore, the least square cubic fit is

$$y = a'_3 x^3 + a'_2 x^2 + a'_1 x + a'_0$$

Example 2.35

Find the exponential equation that is the least square fit to the data points: $(1,1), (2,2), (3,4)$.

Solution

Let $y = ae^{bx}$ be the exponential equation

$$y = ae^{bx}$$

We take the natural log of both sides, we have

$$Iny = bx + Ina$$

$\therefore Iny = bx + Ina$ is the least square exponential curve.

$$In1 = b + Ina$$

$$In2 = 2b + Ina$$

$$In4 = 3b + Ina$$

In matrix form, we have

$$\begin{pmatrix} 1 & 1 \\ 2 & 1 \\ 3 & 1 \end{pmatrix} \begin{pmatrix} b \\ Ina \end{pmatrix} = \begin{pmatrix} 0 \\ In2 \\ 2In2 \end{pmatrix}$$

$$\begin{pmatrix} b' \\ Ina' \end{pmatrix} = \left[\begin{pmatrix} 1 & 2 & 3 \\ 1 & 1 & 1 \end{pmatrix} \begin{pmatrix} 1 & 1 \\ 2 & 1 \\ 3 & 1 \end{pmatrix} \right]^{-1} \begin{pmatrix} 1 & 2 & 3 \\ 1 & 1 & 1 \end{pmatrix} \begin{pmatrix} 0 \\ In2 \\ 2In2 \end{pmatrix}$$

$$\begin{pmatrix} b' \\ Ina' \end{pmatrix} = \begin{pmatrix} 14 & 6 \\ 6 & 3 \end{pmatrix}^{-1} \begin{pmatrix} 1 & 2 & 3 \\ 1 & 1 & 1 \end{pmatrix} \begin{pmatrix} 0 \\ In2 \\ 2In2 \end{pmatrix}$$

$$= \frac{1}{6} \begin{pmatrix} 3 & -6 \\ -6 & 14 \end{pmatrix} \begin{pmatrix} 1 & 2 & 3 \\ 1 & 1 & 1 \end{pmatrix} \begin{pmatrix} 0 \\ In2 \\ 2In2 \end{pmatrix}$$

$$= \frac{1}{6} \begin{pmatrix} -3 & 0 & 3 \\ 8 & 2 & -4 \end{pmatrix} \begin{pmatrix} 0 \\ In2 \\ 2In2 \end{pmatrix}$$

$$= \frac{1}{6} \begin{pmatrix} 6In2 \\ -6In2 \end{pmatrix} = \begin{pmatrix} In2 \\ -In2 \end{pmatrix}$$

$$b' = In2 , Ina' = -In2$$

$$b' = 2 , a' = \frac{1}{2}$$

Therefore, the least square exponential curve is

$$\Rightarrow \quad y = \frac{1}{2}e^{xIn2} = \frac{1}{2} \cdot 2^x$$

$$y = 2^{x-1}$$

Note that, in fitting a least square exponential curve, the assumption that the equation is of the form $y = ae^{bx}$ is not harmful. This is because if

$$y = ac^{bx}, c \neq e, \text{then}$$

$$c = e^{Inc}$$

$$\Rightarrow \quad y = a(e^{Inc})^{bx}$$

$$\Rightarrow \quad y = ae^{(bInc)x}$$

$$\therefore y = ae^{kx}, where \ k = bInc$$

This means that there is no harm is assuming the form $y = ae^{bx}$.

Chapter Three
Linear Transformation

LINEAR TRANSFORMATION

The concept of **Linear Transformation** is a generalization of that of function usually encountered in one-variable calculus. Essentially, a transformation, also called **mapping,** is a function from one vector space U into another vector space V. A transformation, like the usual function, has three elements: two sets, the **domain** and the **codomain**, and a **rule of correspondence.** If the domain of a function is of dimension two and above, we call such a function, mapping or transformation. Therefore, a transformation is a multidimensional function T that associates with each member $\boldsymbol{u}$ of the domain U, a unique member $\boldsymbol{v}$ of the codomain V according to a rule of correspondence. Conventionally, this is denoted as:

$$T: U \rightarrow V.$$

The subset of the co-domain V that contains elements of the form $T(\boldsymbol{u})$ only is called the range of T. Clearly, the range, then, is always a subset of the codomain V. When the range is not a proper subset of V; that is, when the range is the same as V, then the transformation T is said to be **onto.** In other words, an onto transformation T is that in which every element of the codomain V is of the form $T(\boldsymbol{u})$. A transformation T is said to be **one-to-one** if distinct elements $\boldsymbol{u}_1$, $\boldsymbol{u}_2$ in U have distinct transformation values $T(\boldsymbol{u}_1)$, $T(\boldsymbol{u}_2)$ in V. In other words, a transformation T is one-to-one if $T(\boldsymbol{u}_1) = T(\boldsymbol{u}_2)$ in V always implies that $\boldsymbol{u}_1 = \boldsymbol{u}_2$ in U. The transformation value of any $\boldsymbol{u}$ in V, denoted as $T(\boldsymbol{u})$ is called the **image** of $\boldsymbol{u};$ henceforth, we adopt this terminology.

A transformation T: U $\rightarrow$ V is said to be linear if for every $\boldsymbol{u}_1, \boldsymbol{u}_2$ in U, and any real scalar $k,$

* $$T(k\boldsymbol{u}_1) = k\,T(\boldsymbol{u}_1)$$

* $$T(\boldsymbol{u}_1 + \boldsymbol{u}_2) = T(\boldsymbol{u}_1) + T(\boldsymbol{u}_2)$$

These two conditions can be jointly stated as

* $T(k_1\boldsymbol{u}_1 + k_2\boldsymbol{u}_2) = k_1T(\boldsymbol{u}_1) + k_2T(\boldsymbol{u}_2)$ for scalars k_1 and k_2.

Any transformation that satisfies the above condition is said to be a **linear transformation.** The following are examples of linear transformation.

Example 3.1

Stretching

Let U and V both be Euclidean plane and let T be the transformation that transforms each point (u_1, u_2) in U to the point $(2u_1, 2u_2)$. This is called **stretching.** Show that T is a linear transformation.

Solution

We simply need to show that for any $\boldsymbol{u}_1 = (u_{11}, u_{12})$, $\boldsymbol{u}_2 = (u_{21}, u_{22})$ in U,

$$T(k_1\boldsymbol{u}_1 + k_2\boldsymbol{u}_2) = k_1T(\boldsymbol{u}_1) + k_2T(\boldsymbol{u}_2) \text{ for scalars } k_1 \text{ and } k_2.$$

$$k_1\boldsymbol{u}_1 = (k_1u_{11}, k_1u_{12})\,, k_2\boldsymbol{u}_2 = (k_2u_{21}, k_1u_{22})$$

$$T(k_1\boldsymbol{u}_1) = (2k_1u_{11}, 2k_1u_{12}), T(k_2\boldsymbol{u}_2) = (2k_2u_{21}, 2k_2u_{22})$$

$$T(k_1\boldsymbol{u}_1 + k_2\boldsymbol{u}_2) = T(k_1u_{11} + k_2u_{21}, k_1u_{12} + k_2u_{22})$$

$$= (2(k_1u_{11} + k_2u_{21}), 2(k_1u_{12} + k_2u_{22}))$$

$$= (2k_1u_{11} + 2k_2u_{21}, 2k_1u_{12} + 2k_2u_{22})$$

$$T(k_1\boldsymbol{u}_1) + T(k_2\boldsymbol{u}_2) = (2k_1u_{11}, 2k_1u_{12}) + (2k_2u_{21}, 2k_2u_{22})$$

$$= (2k_1u_{11} + 2k_2u_{21}, 2k_1u_{12} + 2k_2u_{22})$$

$$T(k_1\boldsymbol{u}_1 + k_2\boldsymbol{u}_2) = T(k_1\boldsymbol{u}_1) + T(k_2\boldsymbol{u}_2)$$

It is easy to see that $T(k_1\boldsymbol{u}_1) = k_1T(\boldsymbol{u}_1)$, $T(k_2\boldsymbol{u}_2) = k_2T(\boldsymbol{u}_2)$

$$\therefore T(k_1\boldsymbol{u}_1 + k_2\boldsymbol{u}_2) = k_1T(\boldsymbol{u}_1) + k_2T(\boldsymbol{u}_2)$$

This means that stretching is a linear transformation.

Example 3.2

Projection

Let $T: U \rightarrow U$ be such that $T(\boldsymbol{u}) = T(u_1, u_2, u_3) = (u_1, u_2)$. This is projection. Show that it is a linear transformation.

Solution

Let $k\boldsymbol{u} = (ku_1, ku_2, ku_3)$; $T(k\boldsymbol{u}) = (ku_1, ku_2) = kT(\boldsymbol{u})$

$$\Rightarrow T(k\boldsymbol{u}) = kT(\boldsymbol{u})$$

$$Let\ \boldsymbol{u}_1 = (u_{11}, u_{12}, u_{13})\ and\ \boldsymbol{u}_2 = (u_{21}, u_{22}, u_{23})\ be\ in\ U,$$

$$Then\ \boldsymbol{u}_1 + \boldsymbol{u}_2 = (u_{11} + u_{21}, u_{12} + u_{22}, u_{13} + u_{23})$$

$$T(\boldsymbol{u}_1 + \boldsymbol{u}_2) = T(u_{11} + u_{21}, u_{12} + u_{22}, u_{13} + u_{23})$$

$$= (u_{11} + u_{21}, u_{12} + u_{22})$$

$$T(\boldsymbol{u}_1) = T(u_{11}, u_{12}, u_{13}) = (u_{11}, u_{12})$$

$$T(\boldsymbol{u}_1) + T(\boldsymbol{u}_2) = (u_{11}, u_{12}) + (u_{21}, u_{22})$$

$$= (u_{11} + u_{21}, u_{12} + u_{22})$$

$$\therefore T(\boldsymbol{u}_1 + \boldsymbol{u}_2) = T(\boldsymbol{u}_1) + T(\boldsymbol{u}_2)$$

Therefore the projection is a linear transformation.

Example

$T: U \rightarrow V$ such that $T(u_1, u_2) = (u_1 + 1, u_2 + 1)$. Is this a linear transformation?

Solution

$$\boldsymbol{u} = (u_1, u_2);\ k\boldsymbol{u} = (ku_1, ku_2)\ for\ any\ scalar\ k$$

$$T(k\boldsymbol{u}) = (ku_1 + 1, ku_2 + 1)$$

$$But\ kT(\boldsymbol{u}) = k(u_1 + 1, u_2 + 1)$$

$$kT(\boldsymbol{u}) = (ku_1 + k, ku_2 + k)$$

$$\Rightarrow T(k\boldsymbol{u}) \neq kT(\boldsymbol{u})$$

Therefore, T is not a linear transformation.

Example 3.3

$T : U \rightarrow V$ such that $T(u_1, u_2) = (u_1, -2u_2)$. Show that T is a linear transformation.

Solution

$$ku = (ku_1, ku_2); \; T(ku) = (ku_1, -2ku_2) \; for \; any \; scalar \; k.$$
$$kT(u) = k(u_1, -2u_2) = (ku_1, -2ku_2).$$
$$Let \; u_1 = (u_{11}, u_{12}), u_2 = (u_{21}, u_{22}), then$$
$$u_1 + u_2 = (u_{11} + u_{21}, u_{12} + u_{22}).$$
$$T(u_1 + u_2) = T(u_{11} + u_{21}, u_{12} + u_{22}) = (u_{11} + u_{21}, -2u_{12} - 2u_{22})$$
$$But \; T(u_1) = (u_{11}, -2u_{12}), T(u_2) = (u_{21}, -2u_{22})$$
$$T(u_1) + T(u_2) = (u_{11} + u_{21}, -2u_{12} - 2u_{22})$$
$$\Rightarrow \; T(u_1 + u_2) = T(u_1) + T(u_2)$$

Therefore, T is a linear transformation.

Example 3.4

$T : U \rightarrow V$ such that $T(u_1, u_2) = (-u_1, u_1 - u_2, u_2)$. Show that T is a linear transformation.

Solution

As before,

$$ku = (ku_1, ku_2)$$
$$T(ku) = T(ku_1, ku_2) = (-ku_1, ku_1 - ku_2, ku_2)$$
$$= k(-u_1, u_1 - u_2, u_2) = k \, T(u)$$
$$u_1 = (u_{11}, u_{12}), \; u_2 = (u_{21}, u_{22})$$
$$T(u_1) = (-u_{11}, u_{11} - u_{12}, u_{12}), T(u_2) = (-u_{21}, u_{21} - u_{22}, u_{22})$$
$$T(u_1) + T(u_2) = (-u_{11} - u_{21}, u_{11} - u_{12} + u_{21} - u_{22}, u_{12} + u_{22})$$
$$u_1 + u_2 = (u_{11} + u_{21}, u_{12} + u_{22})$$
$$\Rightarrow T(u_1 + u_2) = (-(u_{11} + u_{21}), u_{11} + u_{21} - (u_{12} + u_{22}), u_{12} + u_{22})$$
$$T(u_1 + u_2) = (-u_{11} - u_{21}, u_{11} - u_{12} + u_{21} - u_{22}, u_{12} + u_{22})$$
$$\Rightarrow \; T(u_1) + T(u_2) = T(u_1 + u_2).$$

Therefore, T is a linear transformation.

Example 3.5

Let A be any mxn real matrix. We define a transformation $T: \mathbb{R}^n \to \mathbb{R}^m$ from the vector space of column n-tuples to the space of column m-tuples by $T(X) = A(X)$. Show that T is a linear transformation.

Solution

Let

$$X = \begin{pmatrix} x_1 \\ x_2 \\ \vdots \\ \vdots \\ x_n \end{pmatrix}$$

$$\Rightarrow \; kX = \begin{pmatrix} kx_1 \\ kx_2 \\ \vdots \\ \vdots \\ kx_n \end{pmatrix}$$

$$A(kX) = A \begin{pmatrix} kx_1 \\ kx_2 \\ \vdots \\ \vdots \\ kx_n \end{pmatrix} = kA \begin{pmatrix} x_1 \\ x_2 \\ \vdots \\ \vdots \\ x_n \end{pmatrix} = kA(X)$$

Also, if

$$X_1 = \begin{pmatrix} x_{11} \\ x_{12} \\ \vdots \\ \vdots \\ x_{1n} \end{pmatrix} \text{ and } X_2 = \begin{pmatrix} x_{21} \\ x_{22} \\ \vdots \\ \vdots \\ x_{2n} \end{pmatrix}, \text{ then}$$

$$A(X_1 + X_2) = A \begin{pmatrix} x_{11} + x_{21} \\ x_{12} + x_{22} \\ \vdots \\ \vdots \\ x_{1n} + x_{2n} \end{pmatrix} = A \begin{pmatrix} x_{11} \\ x_{12} \\ \vdots \\ x_{1n} \end{pmatrix} + A \begin{pmatrix} x_{21} \\ x_{22} \\ \vdots \\ x_{2n} \end{pmatrix}$$

$\Rightarrow A(X_1 + X_2) = AX_1 + AX_2$; hence, T is a linear transformation.

The following are left to the reader as exercises.

Example 3.6

Show that the transformation $T: R^3 \rightarrow R^3$ given by $T(x, y, z) = (x, 1, 1)$ is not a linear transformation.

Example 3.7

Show that the transformation $T: R^3 \rightarrow P_2$ from Euclidean 3-space to the vector space of polynomials of degree two or less given by $T(a, b, c) = ax^2 + bx + c$ is a linear transformation.

Example 3.8

The transformation $T: U \rightarrow U$ defined by $T(u) = u$ is a linear transformation. This is called the **identity linear transformation**.

Proof

$$u = (u_1, u_2), ku = (ku_1, ku_2)$$
$$T(ku_1, ku_2) = (ku_1, ku_2) = k(u_1, u_2) = kT(u)$$
$$u_1 = (u_{11}, u_{12}), u_2 = (u_{21}, u_{22})$$
$$\Rightarrow T(u_1) = (u_{11}, u_{12}), T(u_2) = (u_{21}, u_{22})$$
$$u_1 + u_2 = (u_{11} + u_{21}, u_{12} + u_{22})$$
$$\Rightarrow T(u_1 + u_2) = T(u_{11} + u_{21}, u_{12} + u_{22})$$
$$= (u_{11} + u_{21}, u_{12} + u_{22})$$
$$T(u_1) + T(u_2) = (u_{11}, u_{12}) + (u_{21}, u_{22}) = (u_{11} + u_{21}, u_{12} + u_{22})$$
$$\Rightarrow T(u_1) + T(u_2) = T(u_1 + u_2).$$ Therefore, T is a linear transformation.

Example 3.9

The transformation $T: U \rightarrow U$ defined by $T(u) = 0$ for all u in U is a linear transformation. This is called the **zero** linear transformation.

Proof

$$T(k\boldsymbol{u}) = T(ku_1, ku_2) = \boldsymbol{0} = (0,0)$$
$$kT(\boldsymbol{u}) = k(0,0) = (0,0)$$
$$kT(\boldsymbol{u}) = T(k\boldsymbol{u})$$
$$\boldsymbol{u}_1 = (u_{11}, u_{12}), \boldsymbol{u}_2 = (u_{21}, u_{22})$$
$$T(\boldsymbol{u}_1) = T(u_{11}, u_{12}) = (0,0) \,; T(\boldsymbol{u}_2) = (0,0)$$
$$T(\boldsymbol{u}_1) + T(\boldsymbol{u}_2) = (0,0) + (0,0) = (0,0)$$
$$\boldsymbol{u}_1 + \boldsymbol{u}_2 = (u_{11} + u_{21}, u_{12} + u_{22})$$
$$\Rightarrow T(\boldsymbol{u}_1 + \boldsymbol{u}_2) = T(u_{11} + u_{21}, u_{12} + u_{22}) = (0,0)$$
$$\Rightarrow T(\boldsymbol{u}_1 + \boldsymbol{u}_2) = T(\boldsymbol{u}_1) + T(\boldsymbol{u}_2).$$ Therefore, T is a linear transformation.

KERNEL AND RANGE OF A LINEAR TRANSFORMATION

If $T:U \rightarrow V$ is a linear transformation from a vector space U into a vector space V, then $T(\boldsymbol{0}) = \boldsymbol{0}$. This simply states that the zero vector (element) of the domain U always maps to the zero (element) of the co-domain V. This is because

$$\boldsymbol{0} + \boldsymbol{0} = \boldsymbol{0}$$
$$\Rightarrow T(\boldsymbol{0} + \boldsymbol{0}) = T(\boldsymbol{0})$$
$$\Rightarrow T(\boldsymbol{0}) + T(\boldsymbol{0}) = T(\boldsymbol{0})$$
$$\Rightarrow 2T(\boldsymbol{0}) = T(\boldsymbol{0})$$

Since

$$T(\boldsymbol{0}) + \boldsymbol{0} = T(\boldsymbol{0})$$
$$2\,T(\boldsymbol{0}) = T(\boldsymbol{0}) + \boldsymbol{0}$$
$$\Rightarrow 2\,T(\boldsymbol{0}) - T(\boldsymbol{0}) = \boldsymbol{0}$$
$$\Rightarrow T(\boldsymbol{0}) = \boldsymbol{0}$$

It may well happen that for certain vectors $\boldsymbol{u}$ in U, $\boldsymbol{u} \neq \boldsymbol{0}$, $T(\boldsymbol{u}) = \boldsymbol{0}$. The set of all such $\boldsymbol{u}$ in U, such that $T(\boldsymbol{u}) = \boldsymbol{0}$ is called the kernel of the linear transformation T or the **null space** of T.

The null space or kernel of $T : U \rightarrow V$ is always a subspace of U.

To show this, we simply need to show that if u_1, u_2 belong to the kernel of T, then for every scalar k, ku_1 belongs to the kernel of T, and $u_1 + u_2$ also belongs to the kernel of T.

Now if u_1 *and* u_2 are in the null space of T, then $T(u_1) = T(u_2) = 0$

$$T(ku_1) = kT(u_1) \; ; \; T \text{ being a linear transformation.}$$
$$\Rightarrow \quad T(ku_1) = k.\mathbf{0} = \mathbf{0}; \text{ also,}$$
$$T(u_1 + u_2) = T(u_1) + T(u_2); \text{ linearity}$$
$$\Rightarrow T(u_1 + u_2) = \mathbf{0} + \mathbf{0} \; ; \text{ because } u_1 \text{ and } u_2 \text{ are in the kernel of } T.$$
$$\Rightarrow \quad T(u_1 + u_2) = \mathbf{0}$$

This allows us to conclude that the null space of T is a subspace of U.

Example 3.10

Let $T : R^3 \rightarrow R^2$ be defined by $T(x, y, z) = (x - y, x - z)$ for every (x, y, z) in R^3. Determine the null space (kernel) of T.

Solution

This is very easy. The kernel of T is the subset of R^3 such that $T(x, y, z) = (0,0)$

$$\Rightarrow \quad (x - y, x - z) = (0,0)$$
$$\Rightarrow \quad x - y = 0, x - z = 0$$
$$\Rightarrow \quad x = y = z.$$

Therefore, the kernel of T is the set of all vectors of the form (x, x, x) in R^3; that is, all vectors of the form $x(1,1,1)$ where x is a scalar. This further tells us that the null space of T is spanned by the vector $(1,1,1)$, and hence is of dimension one. It is obvious that the range of T, which is the set of all elements of the form $(x - y, x - z)$ in R^2, clearly spans R^2. Therefore, the dimension of the range of T is two. Let the kernel of T be denoted as $ker(T)$, and the range of T, range (T); henceforth, we adopt these notations.

Example 3.11

Determine the kernel of $T:U \rightarrow V$ such that $T(u_1, u_1) = T(-u_1, u_1 - u_2, u_2)$.

Solution

$ker\ (T)$ is every (u_1, u_1) in U such that $T(u_1, u_1) = (0,0,0)$. That is, $ker\ (T)$ is such that $(-u_1, u_1 - u_2, u_2) = (0,0,0)$

$\Rightarrow \quad u_1 = 0,\ u_1 = u_2,\ u_2 = 0$

Therefore, the only member of $ker(T)$ is the zero element $(0,0)$; hence $ker\ (T)$ is of dimension zero. Notice that the dimension of the range of T is two because, in matrix form, the range vector $(-u_1, u_1 - u_2, u_2)$ can be written as

$$\begin{pmatrix} -1 & 0 \\ 1 & -1 \\ 0 & 1 \end{pmatrix} \begin{pmatrix} u_1 \\ u_2 \end{pmatrix}$$

We need to reduce the coefficient matrix into its row-echelon form so as to determine the number of independent vectors that span the range. In other words, the rank of this matrix gives us the dimension of the range of T. Let us go.

$$\begin{pmatrix} -1 & 0 \\ 1 & -1 \\ 0 & 1 \end{pmatrix} (R_2 + R_1) \begin{pmatrix} -1 & 0 \\ 0 & -1 \\ 0 & 1 \end{pmatrix} (R_3 + R_2) \begin{pmatrix} -1 & 0 \\ 0 & -1 \\ 0 & 0 \end{pmatrix} (-R_1),(-R_2) \begin{pmatrix} 1 & 0 \\ 0 & 1 \\ 0 & 0 \end{pmatrix}$$

This means that the vectors $(1,0)$, $(0,1)$ span the range T. Therefore the range of T is of dimension 2.

For the last example, the range of T is $(x - y, x - z)$. In matrix form, we have

$$\begin{pmatrix} 1 & -1 & 0 \\ 1 & 0 & -1 \end{pmatrix} \begin{pmatrix} x \\ y \\ z \end{pmatrix}$$

$\begin{pmatrix} 1 & -1 & 0 \\ 1 & 0 & -1 \end{pmatrix} (R_2 - R_1) \begin{pmatrix} 1 & -1 & 0 \\ 0 & 1 & -1 \end{pmatrix}$,which is in row-echelon form.

Therefore, the dimension of the range is two, and is spanned by the set

$$\{(1, -1, 0), (1, 0, -1)\}.$$

Example 3.12

Consider the projection transformation $T : U \rightarrow V$ such that

$T(u_1, u_2, u_3) = (u_1, u_2)$. Determine the null space of T, and the dimensions of the null space and that of the range of T.

Solution

Clearly, $ker(T)$ is such that $T(u_1, u_2, u_3) = (0,0)$

$$\Rightarrow \quad (u_1, u_2) = (0,0)$$

$$\Rightarrow \quad u_1 = 0 , u_2 = 0$$

Therefore, $ker(T)$ is the set of all elements of the form $(0,0,u_3)$ in U.

$\Rightarrow \quad ker(T) = \{(0,0,u_3)|u_3 \in R\}$. That is, all elements in U of the form $u_3(0,0,1)$ where u_3 is a scalar. This null space is clearly spanned by only the vector $(0,0,1)$, and so it is of dimension one. Now, for the range, we have

$(u_1, u_2) = \begin{pmatrix} 1 & 0 \\ 0 & 1 \end{pmatrix}\begin{pmatrix} u_1 \\ u_2 \end{pmatrix}$, which is already in row-echelon form. Therefore, the range

is of dimension 2, and is spanned by the set $\{(1,0),(0,1)\}$.

At this point, an observant reader should have noticed a pattern in the dimension of the kernel of linear transformation T, and its range: to wit,

The dimension of kernel of T + The dimension of the range of T

= the dimension of the domain of T.

This is not an accident; it is always the case that the dimensions of $ker(T)$ and Range(T) sum up to give the dimension of the domain of T. This is known as **Sylvester's law of nullity.** The dimension of $ker(T)$ is also called the **nullity** of T, while that of range(T) is also called the rank of T. therefore, Sylvester's law of nullity states that:

Nullity of T + Rank of T

= Dimesion of the domain U of T $----P173$

MATRIX REPRESENTATIONS OF A LINEAR TRANSFORMATION

Recall that we considered multiplication of vectors X by the matrix A as a linear transformation. The fact is this: every linear transformation including the ones we have considered can be represented by a matrix. In other words, a linear transformation $T: U \to V$ can be regarded as multiplying the vectors in U by a certain matrix A to obtain the range of T. Without wasting time, let us show how to obtain the matrix of transformation.

Let $T: U \to V$ be a linear transformation, if $S = \{u_1, u_2, \ldots u_n\}$ is a basis of U, then any u in U is written as

$$u = a_1 u_1 + a_2 u_2 + \cdots a_n u_n$$
$$\Rightarrow \quad T(u) = T(a_1 u_1 + a_2 u_2 + \cdots a_n u_n)$$
$$= a_1 T(u_1) + a_2 T(u_2) + \cdots a_n T(u_n) \text{ --- linearity of } T.$$

This means that $\{T(u_1), T(u_2), \ldots T(u_n)\}$ spans Range(T), although it may not be a basis since we are not sure that the set $\{T(u_1), T(u_2), \ldots T(u_n)\}$ is linealy independent.

We are simply trying to say that it suffices to transform the basis of U in order to know the effect of T on U. To obtain the matrix form of a linear transformation, we need only to transform the basis of the domain U according to the transformation rule, and then write them as a column vectors to form a matrix. We illustrate this point with an example.

Example 3.13

Find the matrix representation of the linear transformation of $T: R^3 \to R^2$ such that $T(x, y, z) = (x - y, x - z)$ for every (x, y, z) in U.

Solution

A basis for R^3 is $\{(1,0,0), (0,1,0), (0,0,1)\}$

We simply transform each basis vector as follows

$$T(1,0,0) = (1 - 0, 1 - 0) = (1,1)$$

$$T(0,1,0) = (0-1, 0-0) = (-1,0)$$
$$T(0,0,1) = (0-0, 0-1) = (0,-1)$$

We then write these as columns, we have

$\begin{pmatrix} 1 & -1 & 0 \\ 1 & 0 & -1 \end{pmatrix}$. This is the matrix of transformation. If we wish to transform any element of $\mathbb{R}^3$ we simply left multiply by this matrix. For example

$$T(2,-1,3) = \begin{pmatrix} 1 & -1 & 0 \\ 1 & 0 & -1 \end{pmatrix} \begin{pmatrix} 2 \\ -1 \\ 3 \end{pmatrix}$$
$$= \begin{pmatrix} 1(2) + -1(-1) + 0(3) \\ 1(2) + 0(-1) + -1(3) \end{pmatrix}$$
$$= \begin{pmatrix} 3 \\ -1 \end{pmatrix}$$

$\Rightarrow$ $T(2,-1,3) = (3,-1)$. You can confirm this by directly applying the rule of correspondence. Recall that the basis of any vector space is not unique. This means that the matrix representation of a linear transformation is also not unique since it depends on the basis.

This basis $S = \{(1,0),(0,1)\}$ is called the standard basis, and so the matrix representation of T with respect to standard basis is called the standard matrix representation of T. Henceforth, we denote the standard basis with E.

More generally, let U be an n-dimensional vector space with basis $N = \{u_1, u_2, \ldots u_n\}$ and let V be an m-dimensional vector space with basis $S = \{v_1, v_2, \ldots v_m\}$. Let $T: U \to V$ be a linear transformation, then for each u in U, $u = a_1u_1 + a_2u_2 + \cdots a_nu_n$. Therefore, the $N - coordinatization$ of $u = u_N$ or $[u]_N$ is

$$[u]_N = \begin{pmatrix} a_1 \\ a_2 \\ \vdots \\ a_n \end{pmatrix}$$

Since T is a linear transformation, then
$$T(u) = a_1 T(u_1) + a_2 T(u_2) + \ldots a_n T(u_n)$$

Since $T(\boldsymbol{u}_j)$ is in V for $j = 1, 2, \ldots\ldots n$, therefore each $T(\boldsymbol{u}_j)$ can be uniquely written as

$$T(\boldsymbol{u}_j) = b_{i1}\boldsymbol{v}_1 + b_{i2}\boldsymbol{v}_2 + \ldots\ldots b_{im}\boldsymbol{v}_m$$

$$\Rightarrow T(\boldsymbol{u}_j) = \sum_{j=1}^{m} b_{ij}\boldsymbol{v}_i$$

In expanded form,

$$
\begin{array}{llllll}
T(\boldsymbol{u}_1) = & b_{11}\boldsymbol{v}_1 + & b_{12}\boldsymbol{v}_2 + & \ldots\ldots & b_{1m}\boldsymbol{v}_m \\
T(\boldsymbol{u}_2) = & b_{21}\boldsymbol{v}_1 + & b_{22}\boldsymbol{v}_2 + & \ldots\ldots & b_{2m}\boldsymbol{v}_m \\
\vdots & \vdots & \vdots & \vdots & \vdots \\
\vdots & \vdots & \vdots & \vdots & \vdots \\
T(\boldsymbol{u}_n) = & b_{n1}\boldsymbol{v}_1 + & b_{1n}\boldsymbol{v}_2 + & \ldots\ldots & b_{nm}\boldsymbol{v}_m
\end{array}
$$

We now form the matrix A whose j^{th} column is the $S - coordinatization$ of $T(\boldsymbol{u}_j)$.

$$
\text{That is, } A = \begin{pmatrix}
b_{11} & b_{21} & \ldots\ldots & b_{n1} \\
b_{12} & b_{22} & \ldots\ldots & b_{n2} \\
\vdots & \vdots & \ldots\ldots & \vdots \\
b_{1n} & b_{1n} & \ldots\ldots & b_{nm} \\
\downarrow & \downarrow & & \downarrow \\
[T(u_1)]_s & [T(u_2)]_s & & [T(u_n)]_s
\end{pmatrix}.
$$

In abbreviated notation,

$$A = [T]_S^N.$$

This is the matrix representation of the transformation $T{:}U \rightarrow V$ with respect to the basis N and S respectively. This means that the $S - coordinatization$ of the image $\boldsymbol{v}$ in V of the $N - coordinatization$ of $\boldsymbol{u}$ in U can be obtained by left multiplying $[\boldsymbol{u}]_N$ by $[T]_S^N$; That is,

$$[T(\boldsymbol{u})]_S = [\boldsymbol{v}]_S = [T]_S^N[\boldsymbol{u}]_N \qquad ----- P174$$

Example 3.14

Find the matrix representation of the linear transformation $T: R^2 \rightarrow R^2$ such that $T(x, y) = (x - y, 2x + 3y)$ for each (x, y) in R^2 with respect to the standard basis.

Solution

$$\text{The standard basis is } E = \{(1,0), (0,1)\}$$
$$T(1,0) = (1 - 0, 2(1) + 3(0)) = (1,2)$$
$$T(0,1) = (0 - 1, 2(0) + 3(1)) = (-1, 3)$$

In this case, the domain and the range have the same basis, the standard basis E.

$$\therefore [T]_E^E = \begin{pmatrix} 1 & -1 \\ 2 & 3 \end{pmatrix}$$

Conventionally, when the range and the domain have equal basis, we simply write

$$[T]_E^E = [T]_E$$
$$\Rightarrow [T]_E = \begin{pmatrix} 1 & -1 \\ 2 & 3 \end{pmatrix}$$

Example 3.15

Let $T: P_1 \rightarrow P_2$ be given by $T(p(x)) = x\, p(x)$. Find the matrix representation $[T]_H^G$ of this transformation with respect to the bases

$$G = \{u_1 = 1, u_2 = x\}, \; H = \{v_1 = 1, v_2 = x, v_3 = x^2\}.$$

Then use this matrix to compute $T((p(x))$ for $p(x) = 4x + 1$.

Solution

Let $ax + b = (b, a)$, and $ax^2 + bx + c = (c, b, a)$. Therefore, bases G and H are, respectively,

$$G = \{(1,0), (0,1)\}, H = \{(1,0,0), (0,1,0), (0,0,1)\}.$$

The transformation $T(p_1(x)) = xp_1(x) \Rightarrow T(ax + b) = ax^2 + bx$

$$\Rightarrow T(b, a) = (0, b, a).$$
$$\therefore T(1,0) = (0,1,0), \; T(0,1) = (0,0,1).$$

The next thing is to $H - coordinatize$ each of $(0,1,0)$ and $(0,0,1)$.

Let $(0,1,0) = k_1(1,0,0) + k_2(0,1,0) + k_3(0,0,1)$ for some scalars k_1, k_2, k_3.

$\Rightarrow \quad [(0,1,0)]_H = (k_1, k_2, k_3)$. To find k_1, k_2, k_3, we have

$$k_1 + 0 + 0 = 0 \,,\, 0 + k_2 + 0 = 1 \,,\, 0 + 0 + k_3 = 0$$

$$\Rightarrow \quad k_1 = k_3 = 0, \, k_2 = 1$$

$$\Rightarrow [(0,1,0)]_H = (0,1,0), \text{ similarly,}$$

$$[(0,0,1)]_H = (0,0,1)$$

Note that the basis H is the standard basis E for $\mathbb{R}^3$; so, we don't expect any vector (x, y, z) in $\mathbb{R}^3$ to be different from $[(x, y, z)]_H = [(x, y, z)]_E$

$$\Rightarrow [T]_H^G = \begin{pmatrix} 0 & 0 \\ 1 & 0 \\ 0 & 1 \end{pmatrix}$$

$$\therefore T(4x + 1) = T(1, 4)$$

$$= \begin{pmatrix} 0 & 0 \\ 1 & 0 \\ 0 & 1 \end{pmatrix} \begin{pmatrix} 1 \\ 4 \end{pmatrix} = \begin{pmatrix} 0 \\ 1 \\ 4 \end{pmatrix}$$

$$\Rightarrow T(4x + 1) = (0,1,4) = 4x^2 + x, \text{ as expected}$$

Example 3.16

Let $T{:}P_1 \rightarrow P_2$ be the same as the last example. Show that

a) $u_1 = 1 + x$, $u_2 = 1 - x$ form a basis K for P_1

b)   Show that $v_1 = 1 + x$, $v_2 = 1 - x^2$, $v_3 = 1 - x$ form a basis M for P_2

c)   Find the matrix $[T]_M^K$

d)   Use the results of part (c) to compute $\rightarrow [T(p(x))]_M$ for $p(x) = 4x + 1$

Solution

a) $u_1 = 1 + x \equiv (1,1)$; $u_2 = 1 - x \equiv (1,-1)$

$\begin{pmatrix} 1 & 1 \\ 1 & -1 \end{pmatrix} (R_2 - R_1) \begin{pmatrix} 1 & 1 \\ 0 & -2 \end{pmatrix}$. Since u_1 and u_2 are independent, and the dimension of P_1 is 2; therefore, u_1, u_2 form a basis for P_1.

143

b) $v_1 = 1 + x \equiv (1,1,0); \; v_2 = 1 - x^2 \equiv (1,0,-1); \; v_3 = 1 - x \equiv (1,-1,0)$

$$\begin{pmatrix} 1 & 1 & 0 \\ 1 & 0 & -1 \\ 1 & -1 & 0 \end{pmatrix} (R_3 - R_1) \begin{pmatrix} 1 & 1 & 0 \\ 1 & 0 & -1 \\ 0 & -2 & 0 \end{pmatrix} (R_2 - R_1) \begin{pmatrix} 1 & 1 & 0 \\ 0 & -1 & -1 \\ 0 & -2 & 0 \end{pmatrix}$$

$$(R_3 - 2R_2) \begin{pmatrix} 1 & 1 & 0 \\ 0 & -1 & -1 \\ 0 & 0 & 2 \end{pmatrix}$$

Since the row-echelon form has three nonzero rows, therefore u_1, u_2, and u_3 are linearly independent, and thus form a basis for P_2 because the dimension of P_2 is 3.

c)   We first transform each basis vector to find $[T]_M^K$

$$T(1 + x) = T(1,1) = (0,1,1)$$

$$T(1 - x) = T(1,-1) = (0,1,-1)$$

Next, $k_1(1,1,0) + k_2(1,0,-1) + k_2(1,-1,0) = (0,1,1)$

$$\Rightarrow \quad [(0,1,1)]_M = (k_1, k_2, k_3) \text{ for some scalars } k_1, k_2, k_3$$

$$k_1 + k_2 + k_3 = 0 \, , \, k_1 - k_3 = 1 \, , \, -k_2 + 0 = 1$$

$$\Rightarrow \quad k_2 = -1 \, , \, k_1 = 1 \, , \, k_3 = 0$$

$$\Rightarrow \quad [(0,1,1)]_M = (1,-1,0). \text{ Similarly,}$$

$(0,1,-1) = c_1(1,1,0) + c_2(1,0,-1) + c_3(1,-1,0)$ for some scalars c_1, c_2, c_3.

$$\Rightarrow \quad c_1 + c_2 + c_3 = 0$$

$$c_1 - c_3 = 1$$

$$-c_2 = -1$$

$$\Rightarrow \quad c_2 = 1, c_1 = 0, c_3 = -1$$

$$\Rightarrow \quad [(0,1,-1)]_M = (0,1,-1)$$

$$\therefore [T]_M^K = \begin{pmatrix} 1 & 0 \\ -1 & 1 \\ 0 & -1 \end{pmatrix}$$

d)

$$[T(4x + 1)]_M = [T]_M^K [(4x + 1)]_k$$

$$\Rightarrow [T(1,4)]_M = [T]_M^K [(1,4)]_K$$

$$\text{But } [(1,4)]_K = \begin{pmatrix} a \\ b \end{pmatrix}, where$$

$$(1,4) = a(1,1) + b(1,-1), \text{ for some scalars } a, \text{and } b$$

$$\Rightarrow \quad a + b = 1$$

$$a - b = 4$$

$$\Rightarrow a = \frac{5}{2}, b = \frac{-3}{2}$$

$$\Rightarrow [(1,4)]_K = \frac{1}{2}\begin{pmatrix} 5 \\ -3 \end{pmatrix}$$

Finally,

$$[T(1,4)]_M = \frac{1}{2}\begin{pmatrix} 1 & 0 \\ -1 & 1 \\ 0 & -1 \end{pmatrix}\begin{pmatrix} 5 \\ -3 \end{pmatrix} = \frac{1}{2}\begin{pmatrix} 5 \\ -8 \\ 3 \end{pmatrix}$$

$$\Rightarrow [T(4x+1)]_M = \frac{1}{2}(3x^2 - 8x + 5)$$

THE ALGEBRA OF LINEAR TRANSFORMATION

It has been demonstrated that every linear transformation on a finite dimensional vector space can be thought of as a matrix multiplication. Thus, the algebra of linear transformation is just the same as the algebra of matrices; so, we should not expect something entirely new. Owing to the fact that linear transformations are functions, we can ascribe meaning to the sum, scalar product, and composite (matrix-matrix multiplication) of linear transformation.

Let $T_1: U \rightarrow V$ and $T_2: U \rightarrow V$ be linear transformations from the vector space U into the vector space V, then the sum $T_1 + T_2$ is the function $T_1 + T_2: U \rightarrow V$ defined by $(T_1 + T_2)u = T_1(u) + T_2(u)$ for all u in U. Also, $(kT_1): U \rightarrow V$ is defined by $(kT_1)(u) = kT_1(u) = k(T_1(u)) = T_1(ku)$ for all u in U and for any scalar k.

Suppose that $T_1: U \rightarrow V$ and $T_2: V \rightarrow W$ are linear transformations. The product $T_2T_1: U \rightarrow W$ is defined by $T_2T_1(u) = T_2(T_1(u))$ for all u in U. Just as matrix multiplication, T_2T_1 is usually not equal to T_1T_2 ; that is, in general , $T_2T_1 \neq T_1T_2$.

In fact, $T_1 T_2$ may not even exist (be meaningful) while $T_2 T_1$ does.

Example 3.17

Let $U = V = R^2$. The linear transformations T_1 and T_2 are given by

$$T_1(x_1, x_2) = (-x_1 + 3x_2, x_1 + 2x_2) \text{ and } T_2(x_1, x_2) = (2x_1, x_2 - x_1)$$

Find,

a) $2T_1$

b) $T_1 + 3T_2$

c) $T_1 - T_2$

d) $T_1 T_2$

e) $T_2 T_1$

Solution

a) $(kT_1)(u) = kT_1(u)$ for each u in U, and for any scalar k.

$$\Rightarrow (2T_1)((x_1, x_2)) = 2\, T_1(x_1, x_2)$$

Please note that $T_1((x_1, x_2))$ has been abbreviated as $T_1(x_1, x_2)$. We have been using this simpler notation and we will continue to use it.

$$\Rightarrow (2T_1)((x_1, x_2)) = 2(-x_1 + 3x_2, x_1 + 2x_2)$$
$$= (-2x_1 + 6x_2, 2x_1 + 4x_2)$$

In matrix notation with respect to the standard basis,

$$T_1 = \begin{pmatrix} -1 & 3 \\ 3 & 2 \end{pmatrix} \text{ and } T_2 = \begin{pmatrix} -2 & 0 \\ -1 & 1 \end{pmatrix}$$

$$\therefore \; 2T_1(x_1, x_2) = 2\begin{pmatrix} -1 & 3 \\ 1 & 2 \end{pmatrix}\begin{pmatrix} x_1 \\ x_2 \end{pmatrix} = \begin{pmatrix} -2 & 6 \\ 2 & 4 \end{pmatrix}\begin{pmatrix} x_1 \\ x_2 \end{pmatrix}$$

$$\Rightarrow 2T_1(x_1, x_2) = \begin{pmatrix} -2x_1 + 6x_2 \\ 2x_1 + 4x_2 \end{pmatrix} \text{ or as a row vector,}$$

$$(-2x_1 + 6x_2, 2x_1 + 4x_2).$$

b) $(k_1 T_1 + k_2 T_2)(u)$ is defined by

$$(k_1 T_1 + k_2 T_2)(u) = k_1 T_1(u) + k_2 T_2(u)$$
$$\Rightarrow (T_1 + 3T_2)((x_1, x_2)) = T_1(x_1, x_2) + 3T_2(x_1, x_2)$$
$$= (-x_1 + 3x_2, x_1 + 2x_2) + 3(-2x_1, x_2 - x_1)$$
$$= (-x_1 + 3x_2, x_1 + 2x_2) + (-6x_1, 3x_2 - 3x_1)(-6x1, 3x2 - 3x1)$$
$$= (-7x_1 + 3x_2, -2x_1 + 5x_2)$$

In matrix notation,

$$(T_1 + 3T_2)((x_1, x_2)) = \left[\begin{pmatrix} -1 & 3 \\ 1 & 2 \end{pmatrix} + 3\begin{pmatrix} -2 & 0 \\ -1 & 1 \end{pmatrix}\right]\begin{pmatrix} x_1 \\ x_2 \end{pmatrix}$$

$$= \left[\begin{pmatrix} -1 & 3 \\ 1 & 2 \end{pmatrix} + \begin{pmatrix} -6 & 0 \\ -3 & 3 \end{pmatrix}\right]\begin{pmatrix} x_1 \\ x_2 \end{pmatrix}$$

$$= \begin{pmatrix} -7 & 3 \\ -2 & 5 \end{pmatrix}\begin{pmatrix} x_1 \\ x_2 \end{pmatrix}$$

$$= \begin{pmatrix} -7x_1 + 3x_2 \\ -2x_1 + 5x_2 \end{pmatrix}$$

$$\Rightarrow (7x_1 + 3x_2, -2x_1 + 5x_2)$$

c.) $(T_1 - T_2)((x_1, x_2)) = T_1(x_1, x_2) - T_2(x_1, x_2)$
$$= (-x_1 + 3x_2, x_1 + 2x_2) - (-2x_1, x_2 - x_1)$$
$$= (-x_1 + 3x_2, 2x_1 + x_2)$$

In matrix notation,

$$(T_1 - T_2)((x_1, x_2)) = \left[\begin{pmatrix} -1 & 3 \\ 1 & 2 \end{pmatrix} - \begin{pmatrix} -2 & 0 \\ -1 & 1 \end{pmatrix}\right]\begin{pmatrix} x_1 \\ x_2 \end{pmatrix}$$

$$= \begin{pmatrix} -1 & 3 \\ 2 & 1 \end{pmatrix}\begin{pmatrix} x_1 \\ x_2 \end{pmatrix}$$

$$= \begin{pmatrix} x_1 + 3x_2 \\ 2x_1 + x_2 \end{pmatrix}$$

$$\Rightarrow (x_1 + 3x_2, 2x_1 + x_2)$$

c)
$$(T_1 - T_2)(\boldsymbol{u}) = T_1(T_2(\boldsymbol{u}))$$
$$= T_1(-2x_1, x_2 - x_1)$$
$$= (-(-2x_1) + 3(x_2 - x_1), -2x_1 + 2(x_2 - x_1))$$
$$= (-x_1 + 3x_2, -4x_1 + 2x_2)$$

In matrix notation,
$$(T_1 - T_2)((x_1, x_2)) = \begin{pmatrix} -1 & 3 \\ 1 & 2 \end{pmatrix}\begin{pmatrix} -2 & 0 \\ -1 & 1 \end{pmatrix}\begin{pmatrix} x_1 \\ x_2 \end{pmatrix}$$
$$= \begin{pmatrix} 2-3 & 3 \\ -2-2 & 2 \end{pmatrix}\begin{pmatrix} x_1 \\ x_2 \end{pmatrix}$$

$$\Rightarrow (T_1 T_2)(x_1, x_2) = \begin{pmatrix} -1 & 3 \\ -4 & 2 \end{pmatrix}\begin{pmatrix} x_1 \\ x_2 \end{pmatrix}$$
$$= \begin{pmatrix} -x_1 + 3x_2 \\ -4x_1 + 2x_2 \end{pmatrix}$$
$$\Rightarrow (-x_1 + 3x_2, -4x_1 + 2x_2)$$

d)
$$(T_2 T_1)(x_1, x_2) = T_2(T_1(x_1, x_2))$$
$$= T_2((-x_1 + 3x_2, x_1 + 2x_2))$$
$$= (-2(-x_1 + 3x_2), x_1 + 2x_2 - (-x_1 + 3x_2))$$
$$= (2x_1 - 6x_2, 2x_1 - x_2)$$

In matrix notation,
$$(T_2 T_1)(x_1, x_2) = \begin{pmatrix} -2 & 0 \\ -1 & 1 \end{pmatrix}\begin{pmatrix} -1 & 3 \\ 1 & 2 \end{pmatrix}\begin{pmatrix} x_1 \\ x_2 \end{pmatrix}$$
$$= \begin{pmatrix} 2 & -6 \\ 1+1 & -3+2 \end{pmatrix}\begin{pmatrix} x_1 \\ x_2 \end{pmatrix}$$
$$= \begin{pmatrix} 2 & -6 \\ 2 & -1 \end{pmatrix}\begin{pmatrix} x_1 \\ x_2 \end{pmatrix}$$
$$= \begin{pmatrix} 2x_1 - 6x_2 \\ 2x_1 - x_2 \end{pmatrix}$$
$$\Rightarrow (2x_1 - 6x_2, 2x_1 - x_2)$$

148

Now that we know that a linear transformation on a finite dimensional vector space can always be represented as a matrix, it is not out of place to expect that a linear transformation be **invertible** just as matrices are. Indeed, a linear transformation $T:U \rightarrow V$ is invertible if we find a transformation $S:V \rightarrow U$ such that $ST(u) = u$ for all u in U and $TS(v) = v$ for all v in V.

We then say that T has an inverse S, or S has an inverse T. For simplicity purpose, let us confine the case to that in which the domain and the codomain are each the vector space U; that is, $T:U \rightarrow U$.

Not all linear transformations are invertible since it is not all matrices that are invertible. Generally, it is easier to find the inverse of a linear transformation by inverting its matrix representation than by direct calculation, using the rule of correspondence.

- *The linear transformation $T:U \rightarrow U$ is invertible if and only if any matrix representation A of T is nonsingular. Moreover, the inverse of the linear transformation T, denoted as T^{-1}, is always equal to A^{-1}.*

- Let $T:U \rightarrow U$ be a linear transformation and let $G = \{u_1, u_2 .... u_n)$ be a basis of U. then $T^{-1}: U \rightarrow U$ exists if and only if each (and hence all) of the following equivalent conditions hold:

    i) $\{T(u_1), T(u_2), ... T(u_n)$ is a basis for U
    ii) *Kernel of $T = \{0\}$*
    iii) *Range of $T = U$*

    As in matrix algebra, if T_1 and T_2 are each linear transformation, then the inverse of their composition, if it exists, is $(T_1 T_2)^{-1} = T_2^{-1} T_1^{-1}$.

Example 3.18

Let Let $T_1: R^3 \rightarrow R^3$ and Let $T_2: R^3 \rightarrow R^3$ be defined by

$T_1(x, y, z) = (x - y, z - x, z - y)$, and $T_2(x, y, z) = (z, y, z)$. Find

a) T_1^{-1} with respect to the standard basis E

b.) T_2^{-1} with respect to the standard basis E

Solution

T_1 in matrix form is

$T_1 = \begin{pmatrix} 1 & -1 & 0 \\ -1 & 0 & 1 \\ 0 & -1 & 1 \end{pmatrix}$ to find T_1^{-1}, we use the row- transformation method.

$$\left(\begin{array}{ccc|ccc} 1 & -1 & 0 & 1 & 0 & 0 \\ -1 & 0 & 1 & 0 & 1 & 0 \\ 0 & -1 & 1 & 0 & 0 & 1 \end{array}\right)(R_2+R_1)\left(\begin{array}{ccc|ccc} 1 & -1 & 0 & 1 & 0 & 0 \\ 0 & -1 & 1 & 1 & 1 & 0 \\ 0 & -1 & 1 & 0 & 0 & 1 \end{array}\right)$$

$$(R_1-R_2),(R_3-R_2)\left(\begin{array}{ccc|ccc} 1 & 0 & -1 & 0 & -1 & 0 \\ 0 & -1 & 1 & 1 & 1 & 0 \\ 0 & 0 & 0 & -1 & -1 & 1 \end{array}\right)$$

We don't need to proceed because our effort is going to be fruitless; the process will never end simply because T is not invertible. The row of zeros in the row-echelon form of T reveals that T is nonsingular- meaning that the row vectors of A are linearly dependent. Therefore, T_1^{-1} does not exist. Check that some distinct vectors in $\mathbb{R}^3$ have the same image.

b.) $T_2 = \begin{pmatrix} 0 & 0 & 1 \\ 0 & 1 & 0 \\ 0 & 0 & 1 \end{pmatrix}$. We need not proceed from here because there is a column of

zeroes in this matrix. This means that $|T| = 0$; T is nonsingular.

Recall that $|A^T| = A$ for any nxn matrix A. The column vectors are linearly dependent if and only if the row vectors are.

Example 3.19

Let $T_1: \mathbb{R}^2 \to \mathbb{R}^2$ be a reflection in the y-axis. That is,

$T_1(x, y) = (-x, y)$ for every (x, y) in R^2 and $T_2: R^2 \to R^2$, a reflection in the x-axis. That is ,$T_2(x, y) = (x, -y)$. Find the standard matrix representation of the following.

b) T_1^{-1}

c) T_2^{-1}

d) $(T_1 T_2)^{-1}$

e) $(T_2 T_1)^{-1}$

Solution

a) $T_1 = \begin{pmatrix} -1 & 0 \\ 0 & 1 \end{pmatrix}$ in standard form; that is, with respect to basis E

$$\Rightarrow T_1^{-1} = \begin{pmatrix} -1 & 0 \\ 0 & 1 \end{pmatrix}; \text{ see that } T_1 = T_1^{-1} = T_1^T$$

b)

$$T_2 = \begin{pmatrix} 1 & 0 \\ 0 & -1 \end{pmatrix}$$

$$\Rightarrow T_1^{-1} = \begin{pmatrix} 1 & 0 \\ 0 & -1 \end{pmatrix};$$

Also,

$$T_2 = T_2^{-1} = T_2^T$$

c)

$$(T_1 T_2)^{-1} = T_2^{-1} T_1^{-1} = \begin{pmatrix} 1 & 0 \\ 0 & -1 \end{pmatrix} \begin{pmatrix} -1 & 0 \\ 0 & 1 \end{pmatrix}$$

$$= \begin{pmatrix} -1 & 0 \\ 0 & -1 \end{pmatrix}$$

d)

$$(T_2 T_1)^{-1} = T_1^{-1} T_2^{-1} = \begin{pmatrix} -1 & 0 \\ 0 & 1 \end{pmatrix} \begin{pmatrix} 1 & 0 \\ 0 & -1 \end{pmatrix}$$

$$= \begin{pmatrix} -1 & 0 \\ 0 & -1 \end{pmatrix}$$

Again

$$(T_1 T_2)^{-1} = (T_2 T_1)^{-1}.$$

151

This is expected because

$$(T_1T_2)^{-1} = T_2^{-1}T_1^{-1}$$
$$= T_2T_1$$

Also,

$$(T_2T_1)^{-1} = T_1^{-1}T_2^{-1}$$
$$= T_1T_2$$

but

$$T_1 = -T_2$$
$$\Rightarrow (T_1T_2)^{-1} = T_2T_1$$
$$= (-T_1)(-T_2) = T_1T_2$$
$$= (T_2T_1)^{-1}$$
$$\therefore (T_1T_2)^{-1} = (T_2T_1)^{-1}$$

Or

$$T_1 = T_1^T \; ; \; T_2 = T_1^T$$
$$\Rightarrow \quad (T_1T_2)^{-1} = T_2T_1 = T_2^TT_1^T$$
$$= (T_1T_2)^{-1} \text{ , a property of transposition.}$$

Again, since

$$(T_1T_2)^T = T_1T_2$$
$$\therefore (T_1T_2)^{-1} = (T_2T_1)^{-1}.$$

Any such linear transformation (matrix) T, such that $T^T = T^{-1}$ is called *orthogonal matrix* or *orthogonal transformation*. The following calculation also shows us that the product of orthogonal matrices T_1 and T_2 is also orthogonal.
Let

$$T_1^{-1} = T_1^T \, , \; T_2^{-1} = T_2^T, then \; (T_1T_2)^{-1} = (T_1T_2)^T.$$

Because

$$(T_1T_2)^{-1} = T_2^{-1}T_1^{-1}$$
$$= T_2^TT_1^T$$
$$= (T_1T_2)^T$$

Again, note that any reflection matrix T is *symmetric, orthogonal,* and *self Inverse.* By self inverse, we mean $T = T^{-1}$.

Any 2x2 matrix that is orthogonal is either a rotation matrix or a reflection matrix; That is, any 2x2 matrix that is orthogonal conforms to (can be written as) one of the following two matrices A and B.

$$A = \begin{pmatrix} Cos\theta & -Sin\theta \\ Sin\theta & Cos\theta \end{pmatrix}, B = \begin{pmatrix} Cos\theta & Sin\theta \\ Sin\theta & -Cos\theta \end{pmatrix}$$

Where A is the matrix representation of the counterclockwise rotation of the plane about the origin by angle θ, and B is the matrix representation of the reflection of the plane in the line through the point (0,0) whose angle of inclination is $\frac{\theta}{2}$.

CHANGE OF BASIS AND SIMILARITY

Recall that the matrix representation of a linear transformation varies with the basis under consideration. In this section, we simply want to show the relationship between different matrix representations of the same linear transformation; hence, the concept of **similarity of matrices.**

Let $A = [T]_G$ be the matrix representation of a linear transformation $T{:}U \to U$ with respect to a basis G for U, and let $B = [T]_H$ be the matrix representation of the same transformation T with respect to a different basis H for U. If S is the transition from the basis H to the basis G, then

$$A = SBS^{-1}$$

Simply, two matrices A and B are similarly if there is an invertible matrix S, such that

$$A = SBS^{-1}$$

$$OR$$

$$A = P^{-1}BP$$

All the matrix representations of a given linear transformation are similar.

Example 3.20

Let $T: \mathbb{R}^3 \to \mathbb{R}^3$ be the linear transformation defined by

$T(x, y, z) = (x + z, y + x, y + z)$ for every (x, y, z) in $\mathbb{R}^3$. Find,

a) The matrix representation of T with respect to the standard basis E

b) The matrix representation of T with respect to the basis

$G = \{u_1 = (1, -1, 0) , u_2 = (1, 0, -1) , \quad u_3 = (0, 1, 1)\}$

Solution

$$E = \{(1,0,0), (0,1,0), (0,0,1)\}$$

$$T(1,0,0) = (1,1,0), T(0,1,0) = (0,1,1) , T(0,0,1) = (1,0,1)$$

$$\Rightarrow [T]_E = \begin{pmatrix} 1 & 0 & 1 \\ 1 & 1 & 0 \\ 0 & 1 & 1 \end{pmatrix}$$

b) $T(1, -1, 0) = (1, 0, -1), T(1, 0, 1) = (0, 1, -1), T(0, 1, 1) = (1, 1, 2)$

The next thing is to $G - coordinatize$ each of the image vectors of basis G

$$[T(1, -1, 0)]_G = [(1, 0, -1)]_G = u_2 = (0, 1, 0)$$

$$[(0, 1, -1)]_G = u_2 - u_1 = (-1, 1, 0)$$

$$[(1, 1, 2)]_G = u_1 + 2u_3 = (1, 0, 2)$$

$$\Rightarrow [T]_G^G = [T]_G = \begin{pmatrix} 0 & -1 & 1 \\ 1 & 1 & 0 \\ 0 & 0 & 2 \end{pmatrix}$$

If S is the transition matrix from basis E to basis G, then

$$[T]_G = S[T]_E S^{-1}$$

$$\Rightarrow \quad [T]_E = S^{-1}[T]_G S$$

$$\Rightarrow \quad [T]_E = S^{-1}[T]_G (S^{-1})^{-1}$$

154

This means that S^{-1} is the transition matrix from basis G to basis E. The matrix S is found by $G - coordinatizing$ each vector in the basis E and then writing them as column vectors to form matrix S. Similarly, S^{-1}, the inverse of S, is found by E-coordinatizing each of the basis vectors of basis G and then writing them as column vectors to form matrix S^{-1}. We may find S^{-1} by inverting S if it is easier than coordinatizing. The transition matrix S is also called the **change of basis matrix** from E to G, while S^{-1} is the change of basis matrix from G to E.

To find S, we need to $G - coordinatize$ E, but to find S^{-1} we need to E-coordinatize G. obviously, finding S^{-1} is far easier because we need not do any calculation. Let us E-coordinatize G, we have

$$[(1,-1,0)]_E = (1,-1,0), [(1,0,-1)]_E = (1,0,-1), [(0,1,1)]_E = (0,1,1)$$

$$\Rightarrow [T]_E = \begin{pmatrix} 1 & 1 & 0 \\ -1 & 0 & 1 \\ 0 & -1 & 1 \end{pmatrix} [T]_G \begin{pmatrix} 1 & 1 & 0 \\ -1 & 0 & 1 \\ 0 & -1 & 1 \end{pmatrix}^{-1}$$

$$\Rightarrow [T]_G = \begin{pmatrix} 1 & 1 & 0 \\ -1 & 0 & 1 \\ 0 & -1 & 1 \end{pmatrix}^{-1} [T]_E \begin{pmatrix} 1 & 1 & 0 \\ -1 & 0 & 1 \\ 0 & -1 & 1 \end{pmatrix}$$

$$\Rightarrow [T]_G = \begin{pmatrix} \frac{1}{2} & \frac{-1}{2} & \frac{1}{2} \\ \frac{1}{2} & \frac{1}{2} & \frac{-1}{2} \\ \frac{1}{2} & \frac{1}{2} & \frac{1}{2} \end{pmatrix} [T]_E \begin{pmatrix} 1 & 1 & 0 \\ -1 & 0 & 1 \\ 0 & 1 & 1 \end{pmatrix}$$

That is $[T]_G = S[T]_E S^{-1}$

Check that the columns of S are indeed the $G - coordinatization$ of E.

Again, any two matrices A and B are similar if these is a nonsingular matrix S, such that

$$A = SBS^{-1}$$

$$\Rightarrow \quad AS = SB$$

Example 3.21

Is the matrix $A = \begin{pmatrix} 1 & 0 \\ 0 & 1 \end{pmatrix}$ similar to $B = \begin{pmatrix} -1 & 0 \\ 0 & 1 \end{pmatrix}$?

Solution

If A is similar to B, then there is an S such that

$$AS = SB, \text{ let } S = \begin{pmatrix} w & y \\ x & z \end{pmatrix}, \text{ then}$$

$$\begin{pmatrix} 1 & 0 \\ 0 & -1 \end{pmatrix} \begin{pmatrix} w & y \\ x & z \end{pmatrix} = \begin{pmatrix} w & y \\ x & z \end{pmatrix} \begin{pmatrix} -1 & 0 \\ 0 & 1 \end{pmatrix}$$

$$\Rightarrow \begin{pmatrix} w & y \\ -x & -z \end{pmatrix} = \begin{pmatrix} -w & y \\ -x & z \end{pmatrix},$$

$$\Rightarrow 2w = 0, 2z = 0 ; w = 0, z = 0$$

$$\Rightarrow S = \begin{pmatrix} 0 & y \\ x & 0 \end{pmatrix}$$

For $|S| \neq 0, xy \neq 0, \Rightarrow x \neq 0 \text{ and } y \neq 0$

Any non zero x and y values assure us of the existence of S such that

$AS = SB$. Therefore, the answer is yes, A and B are similar since x and y are arbitrary. The relation of similarity has the following features.

Reflexivity

For every nxn matrix A, A is similar to A. That is, every square matrix is similar to itself. This is because

$$A = IAI^{-1}, \text{ where } I \text{ is the identity matrix.}$$

Symmetry

If A is similar to B, then B is similar to A.

This is because if $A = SBS^{-1}$ for some S such that $|S| \neq 0$, then

$$S^{-1}A = S^{-1} SBS^{-1}$$

$$S^{-1}A = IBS^{-1}$$

$$S^{-1}A S = IBS^{-1}S$$

$$S^{-1}AS = IBI = B$$

$$\Rightarrow \qquad B = S^{-1}A(S^{-1})^{-1}$$

$$\Rightarrow \quad B = PAP^{-1}, where\ P = S^{-1}, since\ |S| \neq 0, |P| \neq 0$$

Therefore B is also similar to A

Transitivity

If A is similar to B, and B is similar to C, then A is similar to C.

$$For\ If\ A = SBS^{-1}\ , and$$

$$B = TCT^{-1}, then$$

$$A = S(TCT^{-1})S^{-1}$$

$$\Rightarrow \quad A = (ST)C(T^{-1}S^{-1})$$

$$A = STC(ST)^{-1} = PCP^{-1}, where\ P = ST$$

Since $|S| \neq 0, |T| \neq 0$, then $|ST| \neq 0$. Therefore, A is similar to C.

Example 3.22

Suppose A and B are n x n invertible matrices, where $n > 1$, and I is the nxn identity matrix. If A and B are similar matrices, which of the following statement must be true?

i) $A - 2I$ and $B - 2I$ are similar matrices

ii) A and B have the same trace

iii) A^{-1} and B^{-1} are similar matrices

Solution

i) $$A = SBS^{-1}\ for\ |S| \neq 0$$

$$\Rightarrow A - 2I = SBS^{-1} - 2I$$

since

$$IS = SI$$

$$I = SIS^{-1}$$

$$\Rightarrow A - 2I = SBS^{-1} - 2SIS^{-1}$$
$$\Rightarrow A - 2I = S(B - 2I)S^{-1}$$

Therefore $A - 2I$ is similar to $B - 2I$.

ii) $$A = SBS^{-1}$$

Check that $tr(AB) = tr(BA)$ for any $n x n$ matrices A and B

$$\Rightarrow tr(A) = tr(SBS^{-1}) = tr((SB)S^{-1})$$
$$= tr(S^{-1}(SB)) = tr(S^{-1}SB) = tr(B)$$

iii) $$A = SBS^{-1}, |S| \neq 0$$
$$\Rightarrow A^{-1} = ((SB)S^{-1})^{-1}$$
$$A^{-1} = (S^{-1})^{-1}(SB)^{-1}$$
$$A^{-1} = SB^{-1}S^{-1}$$

Therefore A^{-1} and B^{-1} are similar. All the three statements are true.

*An equivalence relation is a relation that is reflexive, symmetric and transitive. Therefore, the relation of similarity on square matrices is an **equivalence relation.***

Any equivalence relation on a set partitions the set into mutually (pairwise) disjoint subsets known as equivalence classes such that every object or element of the set belongs to exactly one equivalence class.

All matrices similar to a given n x n matrix A belong to a class called a similarity class. Every n x n matrix belongs to exactly one similarity class, and no two distinct similarity classes have any elements (matrices) in common.

All the possible matrix representations of a given linear transformation are similar, and hence they belong to the same similarity class. Similarity classes on the set of matrices vary with the linear transformation.

ORTHOGONAL LINEAR TRANSFORMATION

Now we know that linear transformation in finite dimensional spaces are matrices. Therefore, an orthogonal linear transformation is simply an orthogonal matrix which is a matrix representation of a linear transformation. One thing that is interesting about orthogonal linear transformation – why we study them – is that they **preserve** inner product, and hence they preserve the length of a vector and the angle between two vectors.

It has been established already, if you recall, that an orthogonal linear transformation (orthogonal matrix) on R^2 conforms to either a rotation of the plane, or a reflection of the plane under the standard inner product. Indeed, the specification or the definition of an inner product function on the vector space under consideration is crucial. This is because the concepts of length and angle preservation are meaningful only when a real inner product has been defined. Thus, the term orthogonality strictly applies to inner product spaces.

To illustrate our claim, let us start with a rotation of the plane. Consider the simple diagram below.

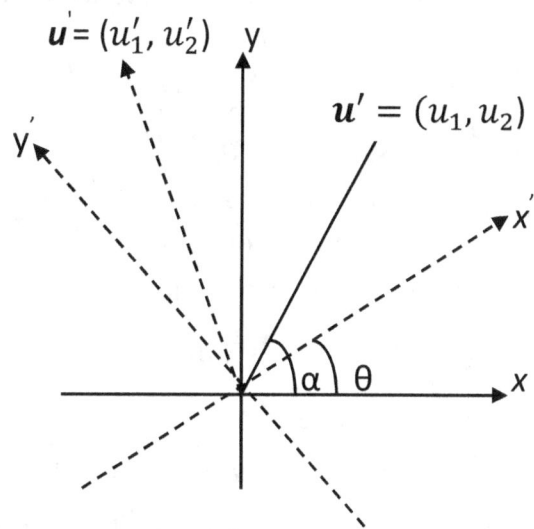

The vector $\boldsymbol{u} = (u_1, u_2)$ initially making an angle α with the x – axis is now at an angle of $(\alpha + \theta)$ with the x – axis, due to a rotation of the plane through an angle θ about the origin.

$$\boldsymbol{u} = (u_1, u_2) \Rightarrow u_1 = |\boldsymbol{u}|\cos\alpha, u_2 = |\boldsymbol{u}|\sin\alpha$$

Let T_θ denote a rotation of the plane through θ about the origin; that is, T_θ is the rotation linear transformation. Therefore

$$T_\theta(\boldsymbol{u}) = T_\theta((u_1, u_2)) = \boldsymbol{u}'$$
$$\Rightarrow \quad T_\theta(\boldsymbol{u}) = \boldsymbol{u}' = (u_1', u_2')$$

Since $\boldsymbol{u}'$ makes an angle of θ +α with the x – axis, therefore

$$u_1' = |\boldsymbol{u}'|\cos(\theta + \alpha), u_2' = |\boldsymbol{u}'|\sin(\theta + \alpha)$$
$$Since \ |\boldsymbol{u}'| = |\boldsymbol{u}|$$
$$u_1' = |\boldsymbol{u}|\cos(\theta + \alpha), u_2' = |\boldsymbol{u}|\sin(\theta + \alpha)$$

$$\Rightarrow u_1' = |\boldsymbol{u}|\cos\theta\cos\alpha - |\boldsymbol{u}|\sin\theta\sin\alpha = u_1\cos\theta - u_2\sin\theta$$

$$u_2' = |\boldsymbol{u}|\sin\theta\cos\alpha + |\boldsymbol{u}|\cos\theta\sin\alpha = u_1\sin\theta + u_2\cos\theta$$
$$\Rightarrow \quad T_\theta((u_1, u_2)) = (u_1\cos\theta - u_2\sin\theta, \ u_1\sin\theta + u_2\cos\theta)$$

This is the linear transformation for rotation of the plane through an angle θ.

In matrix rotation, as usual, T_θ with respect to the standard basis E for $\mathbb{R}^2$, is written as:

$$T_\theta = \begin{pmatrix} \cos\theta & -\sin\theta \\ \sin\theta & \cos\theta \end{pmatrix}$$
$$\Rightarrow T_\theta(u_1, u_2) = \begin{pmatrix} \cos\theta & -\sin\theta \\ \sin\theta & \cos\theta \end{pmatrix}\begin{pmatrix} u_1 \\ u_2 \end{pmatrix}$$

Let us show that T_θ, indeed, preserves inner product.

Let $\boldsymbol{u} = (u_1, u_2) \ and \ \boldsymbol{v} = (v_1, v_2)$ be in $\mathbb{R}^2$, then

$$\langle \boldsymbol{u}, \boldsymbol{v} \rangle = u_1 v_1 + u_2 v_2$$
$$T_\theta(\boldsymbol{u}) = T_\theta(u_1, u_2) = (u_1\cos\theta - u_2\sin\theta, u_1\sin\theta + u_2\cos\theta)$$
$$T_\theta(\boldsymbol{v}) = T_\theta(v_1, v_2) = (v_1\cos\theta - v_2\sin\theta, v_1\sin\theta + v_2\cos\theta)$$

$$\Rightarrow \langle T_\theta(\boldsymbol{u}), T_\theta(\boldsymbol{v}) \rangle$$

$$= (u_1 cos\theta - u_2 sin\theta)(v_1 cos\theta - v_2 sin\theta) + (u_1 sin\theta + u_2 cos\theta)(v_1 sin\theta + v_2 cos\theta)$$

$$= u_1 v_1 cos^2\theta - u_1 v_2 cos\theta sin\theta - u_2 v_1 cos\theta sin\theta + u_2 v_2 sin^2\theta + u_1 v_1 sin^2\theta$$

$$+ u_1 v_2 cos\theta sin\theta + u_2 v_1 cos\theta sin\theta + u_2 v_2 cos^2\theta$$

$$= u_1 v_1 cos^2\theta + u_2 v_2 sin^2\theta + u_1 v_1 sin^2\theta + u_2 v_2 cos^2\theta$$

$$= u_1 v_1 (cos^2\theta + sin^2\theta) + u_2 v_2 (cos^2\theta + sin^2\theta)$$

$$= u_1 v_1 + u_2 v_2$$

$$\Rightarrow \langle \boldsymbol{u}, \boldsymbol{v} \rangle = \langle T_\theta(\boldsymbol{u}), T_\theta(\boldsymbol{v}) \rangle$$

This means that T_θ preserves inner products. This is the definition of an orthogonal linear transformation.

An orthogonal linear transformation T preserves inner product in the sense that for any $\boldsymbol{u}, \boldsymbol{v}$ in the domain of T, $\langle \boldsymbol{u}, \boldsymbol{v} \rangle = \langle T(\boldsymbol{u}), T(\boldsymbol{v}) \rangle$.

The matrix representation with respect to an orthonormal basis of an orthogonal linear transformation T is an orthogonal matrix. That is, If A is the matrix representation of an orthogonal linear transformation T with respect to an orthonormal basis, then $A^T = A^{-1}$.

The converse is also true. That means that a simpler way of verifying that T_θ is an orthogonal transformation is by showing that the matrix A, with respect to the basis E (an orthonormal basis) given by

$A = \begin{pmatrix} cos\theta & sin\theta \\ sin\theta & -cos\theta \end{pmatrix}$, is such that $A^T = A^{-1}$. Indeed it is obvious that such

exactly is the case.

Recall also that the reflection linear transformation, as we saw it, is an orthogonal linear transformation.

Example 3.23

Let $T: \mathrm{R}^2 \rightarrow \mathrm{R}^2$ be such that $T(x_1, x_2) = (-x_2, x_1)$ show that T is an orthogonal linear transformation.

Solution

Obviously, T is a linear transformation. We can confirm its orthogonality in two ways: by direct calculation, as we did for T_θ, and by representing T with respect to an orthonormal basis in matrix form and then testing

If $A^T = A^{-1}$, where A is the matrix representation.

By direct calculation: Let $\boldsymbol{u} = (u_1, u_2)$ and $\boldsymbol{v} = (v_1, v_2)$ in $\mathbb{R}^2$, then

$$\langle \boldsymbol{u}, \boldsymbol{v} \rangle = u_1 v_1 + u_2 v_2$$

$$T(\boldsymbol{u}) = T(u_1, u_2) = (-u_2, u_1)$$

$$T(\boldsymbol{v}) = T(v_1, v_2) = (-v_2, v_1)$$

$$\Rightarrow \langle T(\boldsymbol{u}), T(\boldsymbol{v}) \rangle = (-u_2)(-v_2) + u_1 v_1$$

$$= u_1 v_1 + u_2 v_2$$

$\Rightarrow \langle \boldsymbol{u}, \boldsymbol{v} \rangle = \langle T(\boldsymbol{u}), T(\boldsymbol{v}) \rangle$. Therefore, T is orthogonal linear transformation.

Method 2: The standard basis $E = \{(1,0), (0,1)\}$ is orthonormal. We therefore represent T in matrix form with respect to E. Note that any other orthonormal basis will give the same result, but E is the simplest, of all the orthonormal bases, to deal with.

$$\Rightarrow \quad [T]_E = \begin{pmatrix} 0 & -1 \\ 1 & 0 \end{pmatrix} = A$$

Observe that $A^T = A^{-1}$. Therefore, T is an orthogonal linear transformation.

Let U be an linear product space, and let $T : U \to U$ be a linear transformation. Then T is orthogonal if and only if T preserves length in the sense that $|T(\boldsymbol{u})| = |\boldsymbol{u}|$ for all in $\boldsymbol{u}$ in U.

To show this, suppose that T is orthogonal; that is,

$$\langle \boldsymbol{u}, \boldsymbol{v} \rangle = \langle T(\boldsymbol{u}), T(\boldsymbol{v}) \rangle \text{ for all } \boldsymbol{u}, \boldsymbol{v} \text{ in U.}$$

$$\Rightarrow \langle \boldsymbol{u}, \boldsymbol{u} \rangle^{\frac{1}{2}} = \langle T(\boldsymbol{u}), T(\boldsymbol{u}) \rangle^{\frac{1}{2}}$$

But

$$|\boldsymbol{u}| = \langle \boldsymbol{u}, \boldsymbol{u} \rangle^{\frac{1}{2}}$$

$$\Rightarrow |u| = |T(u)|$$

Now suppose that T preserves length; that is,

$$|u| = |T(u)| \text{ for all } u \text{ in U, then}$$

$$|u + v| = |T(u + v)| \text{ for all } u, v \text{ in U, since } u + v \text{ is also in U.}$$

$$\langle u + v, u + v \rangle^{\frac{1}{2}} = \langle T(u + v), T(u + v) \rangle^{\frac{1}{2}}$$

$$\Rightarrow \langle u, u \rangle + 2\langle u, v \rangle + \langle v, v \rangle = \langle T(u) + T(v), T(u) + T(v) \rangle$$

$$= \langle T(u), T(u) \rangle + 2\langle T(u), T(v) \rangle + \langle T(v), T(v) \rangle$$

Since T preserves length, then

$$\langle u, u \rangle = \langle T(u), T(u) \rangle \text{ and } \langle v, v \rangle = \langle T(v), T(v) \rangle$$

$$\Rightarrow \langle u, v \rangle = \langle T(u), T(v) \rangle.$$

Therefore, T preserves inner product if and only if T preserves length.

Example 3.24

Show that an orthogonal linear transformation preserves angle.

Solution

Recall that for an arbitrary real inner product space U,

$Cos\theta = \dfrac{\langle u,v \rangle}{|u||v|}$ for all u, v in U, where θ is the angle between u and v, $0 \le \theta \le \pi$.

Let α be the angle between $T(u)$ and $T(v)$, then

$$\cos \alpha = \frac{\langle T(u),T(v) \rangle}{|T(u)||T(v)|}$$

Since T preserves both inner product and length, then

$$cos\alpha = \frac{\langle u, v \rangle}{|u||v|}$$

$$\Rightarrow cos\alpha = cos\theta \,; 0 \le \theta, \alpha \le \pi$$

$$\Rightarrow \alpha = \theta$$

Therefore T also preserves angle.

Let $S = \{v_1, v_2, \ldots \ldots v_n\}$ be a basis for the real inner product space V, then the linear transformation $T : V \to V$ is orthogonal if and only if $\langle T(v_i), T(v_j) \rangle = \langle v_i, v_j \rangle$ for each pair of vectors v_i, v_j in S.

To show this, suppose that T is orthogonal, then

$$\langle T(u), T(v) \rangle = \langle u, v \rangle \text{ for any } u, v, \text{ in V}$$

Since S is a basis for V, then

$$u = a_1 v_1 + a_2 v_2 + \ldots a_n v_n \text{ for some scalars } a_i, 1 \le i \le n$$
$$v = b_1 v_1 + b_2 v_2 + \ldots b_n v_n \text{ for some scalars } b_i, 1 \le i \le n$$

$$T(u) = a_1 T(v_1) + a_2 T(v_2) + a_n T(v_n) \text{, being a linear transformation}$$
$$T(v) = b_1 T(v_1) + b_2 T(v_2) + b_n T(v_n)$$
$$\langle T(u), T(v) \rangle = \sum_{i=1}^{n} \sum_{j=1}^{n} a_i b_i \langle T(v_i), T(v_j) \rangle$$

Also,

$$\langle u, v \rangle = \sum_{i=1}^{n} \sum_{j=1}^{n} a_i b_i \langle v_i, v_j \rangle$$

Since

$$\langle v_i, v_j \rangle = \langle T(v_i), T(v_j) \rangle$$

Then

$$\langle T(u), T(v) \rangle = \langle u, v \rangle$$

Now suppose that

$$\langle T(v_i), T(v_j) \rangle = \langle v_i, v_j \rangle.$$

Again, for any u, v in U,

$$u = \sum_{i=1}^{n} a_i v_i \quad v = \sum_{i=1}^{n} b_i v_i$$

$$\langle u, v \rangle = \sum_{i=1}^{n} \sum_{j=1}^{n} a_i b_i \langle v_i, v_j \rangle = \sum_{i=1}^{n} \sum_{j=1}^{n} a_i b_i \langle T(v_i), T(v_j) \rangle$$

But

$$T(u) = \sum_{j=1}^{n} a_i T(v_i) \quad , \quad T(v) = \sum_{i=1}^{n} b_i T(v_i)$$

$$\langle T(u), T(v) \rangle = \sum_{i=1}^{n} \sum_{j=1}^{n} a_i b_i \langle T(v_i), T(v_j) \rangle$$

$$\Rightarrow \langle u, v \rangle = \langle T(u), T(v) \rangle$$

Therefore the fact is proved.

The column vectors of an orthogonal matrix are pairwise orthonormal.

The row vectors of an orthogonal matrix are pairwise orthonormal if and only if the column vectors are pairwise orthonormal.

Let us show the first of these two facts. The second is left to the reader as an exercise.

Let A be an orthogonal matrix, then

$$A^{-1} = A^T$$

But

$$AA^{-1} = I_n = A^{-1}A$$

where A is an $n \times n$ matrix

$$\Rightarrow AA^T = I_n = A^T A$$

Let $c_1, c_2, \dots c_n$ be the column vectors of A, then

$$A = (c_1 : c_2 : \dots c_n)$$

$$\Rightarrow A^T = \begin{pmatrix} c_1^T \\ c_2^T \\ c_n^T \end{pmatrix}$$

$$\Rightarrow A^T A = \begin{pmatrix} c_1^T \\ c_2^T \\ c_n^T \end{pmatrix} (c_1 : c_2 : \ldots c_n)$$

$$\Rightarrow A^T A = \begin{pmatrix} c_1^T c_1 & c_1^T c_2 & \text{-------} & c_1^T c_n \\ c_2^T c_1 & c_2^T c_2 & \text{-------} & c_2^T c_n \\ \vdots & \vdots & & \vdots \\ c_n^T c_1 & c_n^T c_2 & \text{-------} & c_n^T c_n \end{pmatrix}$$

Since

$$A^T A = I_n$$

$$\Rightarrow c_i^T c_i = 1 , c_i^T c_j = 0 \; for \; i \neq j$$

$$\Rightarrow \langle c_i, c_j \rangle = 0 \; for \; i \neq j$$

Therefore, the column vectors are pairwise orthogonal.

Again,

$$c_i^T c_i = 1 \Rightarrow \langle c_i, c_i \rangle = 1 , \Rightarrow |c_i| = 1.$$

This means that each column vector is of unity length. Therefore, the column vectors are pairwise orthonormal.

Chapter Four

An Application To Optics

In this section, we apply matrix algebra as a tool to find out the path of a light ray as it moves through an optical system comprising of lenses and mirrors. In particular we focus on geometric optics, which is the analysis of light in ray approximations.

When a ray of light moves through an optical system, it, at a point in time, either undergoes reflection or refraction, or travels in a straight line in a homogeneous medium bounded between optical surfaces (translation). Upon reflection or refraction at an optical surface, a light ray changes its direction but maintains the same vertical distance from the optical axis immediately before and after reflection or refraction. In other words, vertical distance of a light ray upon reflection or refraction does not change instantaneously. On the other hand, when a ray of light translates (moves freely in a homogenous medium) between two optical surfaces, it does not change its direction; it continues to move in a straight line, but its vertical distance (height) from the optical axis may change. If we assume that the angle made between a light ray and the optical axis is very small (paraxial approximation), then the process of translation, reflection, and refraction of a light ray can be represented by a 2x2 matrix which links the initial and the final states of the light ray. This means that light ray translation, reflection and refraction in paraxial approximation are nothing but linear transformations. Therefore, a sequential multiplication of all the matrix representations of these linear transformations of the whole optical system gives us the overall 2x2 matrix (linear transformation) called the **system matrix**, from which we can deduce all the important features of the composite optical system.

RAY COORDINATES

The coordinate of ray vector are: its normal (perpendicular) distance y from the optic axis, and the angle α which the ray line makes with the optic axis. However, it is more useful to consider a point not as (α, y), but as $(n\alpha, y)$, where n is the refractive index of the medium in which the ray moves. We consider $n\alpha$ and not just α because α is just a geometric angle while $n\alpha$ is the optical angle. We therefore define the state of a light ray at a point in medium of reflective index n by a 2 x 1 column vector R given by

$$R \equiv \begin{pmatrix} n\alpha \\ y \end{pmatrix}$$

TRANSLATION MATRIX

Consider a ray moving in a homogeneous medium of refractive index n between two points 1 and 2 as shown below

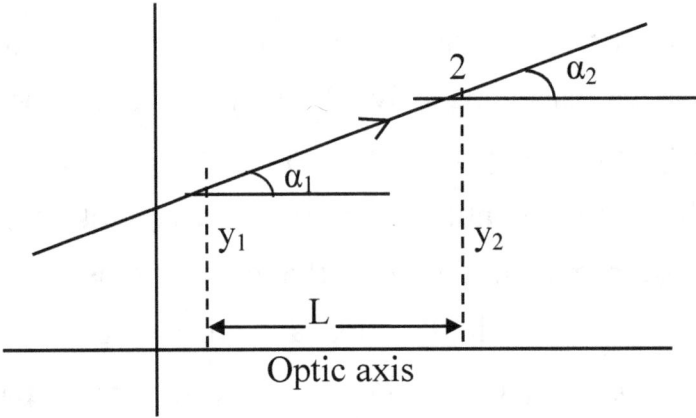

The initial point of the ray is $\begin{pmatrix} n\alpha_1 \\ y_1 \end{pmatrix}$ and the final point is $\begin{pmatrix} n\alpha_2 \\ y_2 \end{pmatrix}$.

That is,

$$R_1 = \begin{pmatrix} n\alpha_1 \\ y_1 \end{pmatrix}; \ R_2 = \begin{pmatrix} n\alpha_2 \\ y_2 \end{pmatrix}.$$

If T is the linear transformation of translation, then

$$T(R_1) = R_2, \qquad \Rightarrow T\begin{pmatrix} n\alpha_1 \\ y_1 \end{pmatrix} = \begin{pmatrix} n\alpha_2 \\ y_2 \end{pmatrix}$$

From the diagram above, it is obvious that

$\alpha_1 = \alpha_2$; $tan\alpha_2 = \frac{y_2 - y_1}{L}$ because the ray is assumed to be paraxial,

$$|\alpha_1| << 1. \text{ Therefore, } tan\alpha_1 \approx \alpha_1$$
$$\Rightarrow \alpha_1 = \alpha_2$$
$$y_2 = L\alpha_1 + y_1.$$
$$\Rightarrow T\begin{pmatrix} n\alpha_1 \\ y_1 \end{pmatrix} = \begin{pmatrix} n\alpha_1 \\ L\alpha_1 + y_1 \end{pmatrix} = \begin{pmatrix} n\alpha_1 \\ L\alpha_1 + y_1 \end{pmatrix} = \begin{pmatrix} n\alpha_1 \\ \frac{L}{n}(n\alpha_1) + y_1 \end{pmatrix}.$$

The matrix representation of T relative to the standard basis E is

$T = \begin{pmatrix} 1 & 0 \\ \frac{L}{n} & 1 \end{pmatrix}$ where L is the horizontal distance between the initial and the find

state of the ray. Note that we rearranged the term $L\alpha_1$ as $\frac{L}{n}(n\alpha_1)$ because the

coordinate is $n\alpha_1$ and not α_1. Again,

$$T\begin{pmatrix} n\alpha_1 \\ y_1 \end{pmatrix} = \begin{pmatrix} 1 & 0 \\ \frac{L}{n} & 1 \end{pmatrix}\begin{pmatrix} n\alpha_1 \\ y_1 \end{pmatrix}$$

$$\Rightarrow \begin{pmatrix} n\alpha_2 \\ y_2 \end{pmatrix} = \begin{pmatrix} 1 & 0 \\ \frac{L}{n} & 1 \end{pmatrix}\begin{pmatrix} n\alpha_1 \\ y_1 \end{pmatrix}$$

The translation matrix is T given by

$$T = \begin{pmatrix} 1 & 0 \\ \frac{L}{n} & 1 \end{pmatrix} \qquad\qquad ------P175$$

Note that

$$|T| = 1$$

REFLECTION MATRIX

Let us consider the reflection of a ray at a convex spherical surface as shown below.

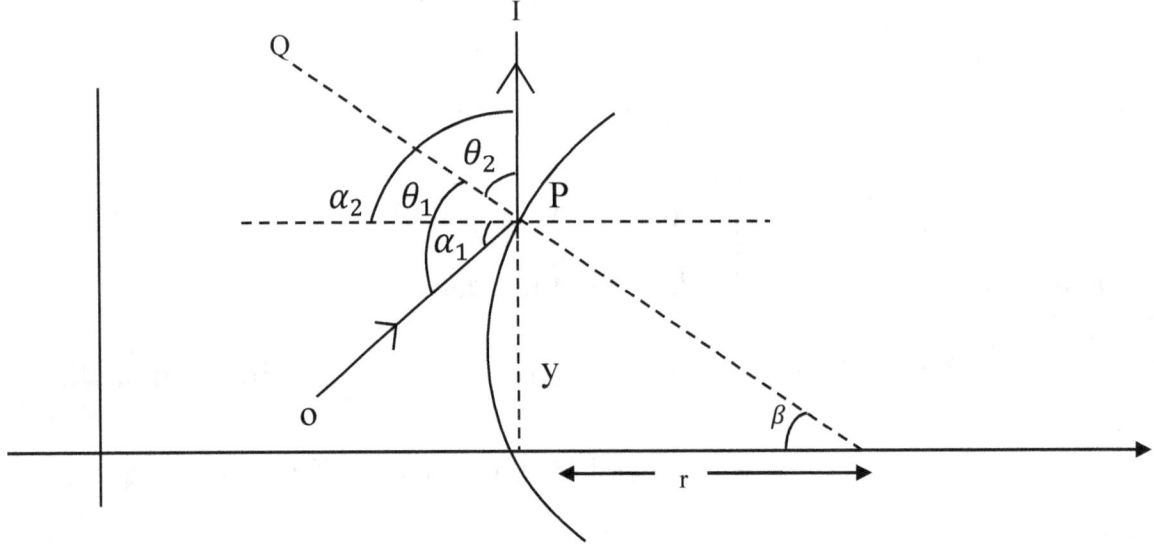

In the diagram, C is the centre of curvature of the spherical surface, and r is its radius of curvature. The point O is the initial point of the light ray – OP is the incident ray and PI is the reflected ray. I is the final state (point) of the ray. θ_1 and θ_2 are the angles of incidence and reflection respectively.

Recall, from basic physics, that the angle of incidence is equal to that of reflection. The line QC is the normal which, of course, passes through C. α_1 and α_2 remain as previously defined. Therefore, from the diagram, we have

$$tan\beta \approx \beta \ for \ |\beta| << 1.$$
$$\Rightarrow \beta = \frac{y}{r}$$
$$\theta_1 = \theta_2 \ ; \ law \ of \ reflection$$
$$\theta_1 = \alpha_1 + \beta$$
$$\alpha_2 = \theta_2 + \beta$$
$$\Rightarrow \alpha_1 + \beta = \alpha_2 - \beta$$

$$\Rightarrow \quad \alpha_2 = \alpha_1 + 2\beta$$

$$\Rightarrow \quad \alpha_2 = \alpha_1 + \frac{2y}{r}$$

Clearly,

$$y_1 = y_2 = y$$

Therefore

$$n\alpha_2 = n\alpha_1 + \frac{2ny_1}{r}$$

$$y_2 = y_1$$

If R is the linear transformation of reflection, then

$$R\begin{pmatrix} n\alpha_1 \\ y_1 \end{pmatrix} = \begin{pmatrix} n\alpha_2 \\ y_2 \end{pmatrix} = \begin{pmatrix} n\alpha_1 + \dfrac{2ny_1}{r} \\ y_1 \end{pmatrix}$$

Again, in matrix form, with respect to the standard basis E,

$R = \begin{pmatrix} 1 & \frac{2n}{r} \\ 0 & 1 \end{pmatrix}$. This is the reflection matrix.

$$\Rightarrow \quad \begin{pmatrix} n\alpha_2 \\ y_2 \end{pmatrix} = \begin{pmatrix} 1 & \frac{2n}{r} \\ 0 & 1 \end{pmatrix} \begin{pmatrix} n\alpha_1 \\ y_1 \end{pmatrix}$$

$$R = \begin{pmatrix} 1 & \frac{2n}{r} \\ 0 & 1 \end{pmatrix} \qquad\qquad ----- P176$$

Note that

$$|R| = 1$$

We define a new parameter f, the focal length. Let f be such that $f = \frac{r}{2}$, then

$$R = \begin{pmatrix} 1 & \frac{n}{f} \\ 0 & 1 \end{pmatrix}$$

REFRACTION MATRIX

Finally, we consider the refraction of a light ray at a spherical surface separating two media of refractive indices n_1 and n_2 as shown below

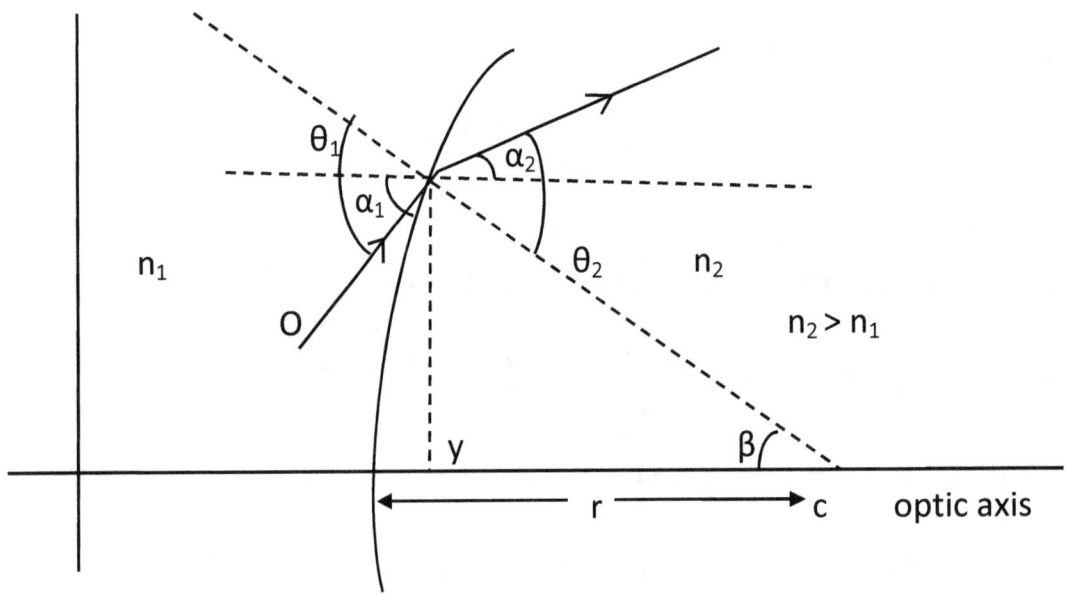

All the parameters, except θ_2, remain as defined for reflection. Here, θ_2 is the angle of refraction.

$\frac{Sin\theta_1}{Sin\theta_2} = \frac{n_2}{n_1}$, snell's law (the law of refraction)

For paraxial rays,

$$Sin\theta_1 \approx \theta_1, \qquad Sin\theta_2 \approx \theta_2$$

$$\Rightarrow \frac{\theta_1}{\theta_2} = \frac{n_2}{n_1}$$

$$tan\,\beta \approx \beta = \frac{y}{r} = \frac{y_1}{r}$$

$$\theta_1 = \alpha_1 + \beta, \; \theta_2 = \alpha_2 + \beta$$

$$\Rightarrow \frac{\theta_1}{\theta_2} = \frac{\alpha_1 + \beta}{\alpha_2 + \beta} = \frac{n_2}{n_1}$$

$$\Rightarrow n_1\alpha_1 + n_1\beta = n_2\alpha_2 + n_2\beta$$

$$\Rightarrow \quad n_2\alpha_2 = n_1\alpha_1 + (n_1 - n_2)\frac{y_1}{r}$$

Let R_f be the refraction linear transformation, then

$$R_f\begin{pmatrix} n\alpha_1 \\ y_1 \end{pmatrix} = \begin{pmatrix} n_2\alpha_2 \\ y_2 \end{pmatrix} = \begin{pmatrix} n_1\alpha_1 + (n_1 - n_2)\frac{y_1}{r} \\ y_1 \end{pmatrix}$$

Therefore R_f with respect to the standard basis E, is given by

$$R_f = \begin{pmatrix} 1 & \frac{n_1 - n_2}{r} \\ 0 & 1 \end{pmatrix}.$$ This is the refraction matrix.

Note that

$$|R_f| = 1$$

$$\Rightarrow \begin{pmatrix} n_2\alpha_2 \\ y_2 \end{pmatrix} = \begin{pmatrix} 1 & \frac{n_1 - n_2}{r} \\ 0 & 1 \end{pmatrix} \begin{pmatrix} n_1\alpha_1 \\ y_1 \end{pmatrix}$$

$$R_f = \begin{pmatrix} 1 & \frac{n_1 - n_2}{r} \\ 0 & 1 \end{pmatrix} \qquad ----- P177$$

Again, we define a new parameter P called the optical power of the refracting surface.

$$P = \frac{n_2 - n_1}{r}.$$

This implies that

$$R_f = \begin{pmatrix} 1 & -P \\ 0 & 1 \end{pmatrix}$$

SIGN CONVENTION

We assumed a convex spherical surface to arrive at our reflection and refraction matrices; also, we assumed that $n_2 > n_1$ in the case of refraction. These, however, do not limit the application of our derivations. We need only adjust the results appropriately as we come across different situations by simply following the sign conventions stated below.

i)   The source point is always taken to the left of the optical system so that incident ray moves from left to right.

ii) All the distances measured to the right of the optical centre are considered positive and those to its left, negative. This implies that the object distance is always negative.

iii) Radius of Curvature, r, is positive for convex surfaces because the centre of curvature c lies to the right of the optical centre. The radius of curvature is negative for concave surfaces.

iv) The angular coordinate α is taken as positive if a ray is pointing upward, and negative if it is pointing downwards. This holds for both initial and final rays.

v)   All the distances measured above the optics axis are taken positive and those below, negative.

SYSTEM MATRIX

A composite optical system consists of several thick and/or thin lenses placed at appropriate separation. We assume that all the component of the system have cylindrical symmetry and are symmetrically located along a common optic axis. Indeed, the propagation of a ray through the composite optical system is equivalent to the product of 2x2 matrices of refraction at various optical surfaces and also translation matrices between the surfaces. The product of all the matrices, in the reverse order, starting from the input surface (first surface) to the output surface (the last surface) gives the **system matrix**. For instance, let us consider the thick lens below.

The thick Lens Matrix: Refraction at two surfaces.

A lens is a device with two refracting surfaces. When a ray of light passes through a lens, refraction occurs at the front surface (input surface). If the lens is considered thick, then the light ray, after refraction, translates i.e. moves freely

174

between the two surfaces in a homogeneous medium, and finally strikes the rear surface where it undergoes a second refraction.

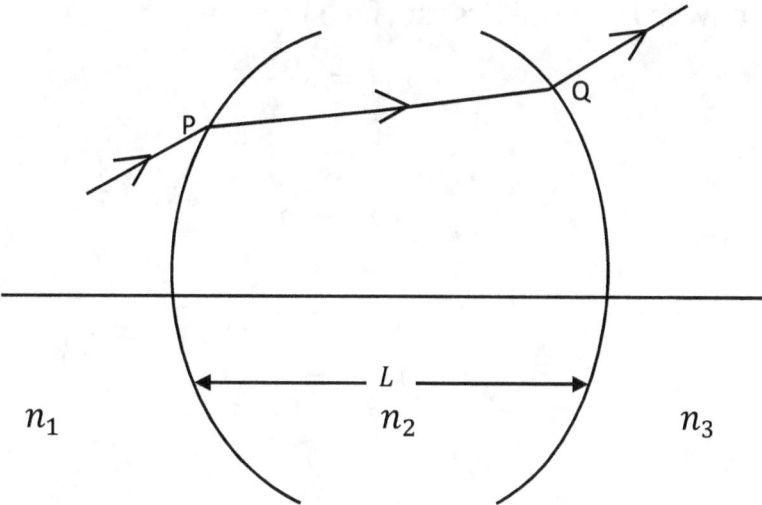

In the above diagram, the thick lens has a thickness of length L. It has two surfaces passing through the points P and Q respectively. It undergoes the first refraction at point P on the first surface. Thereafter, it translates between the surfaces, which are separated by a distance of L. Finally, at point Q, it undergoes a second refraction. The respective refractive indices of the media are also shown. Let R_{fP}, T, and R_{fQ}, be the respective matrices of refraction at P, translation from P to Q, and refraction at Q, then the system matrix S is given by

$$S = R_{fQ}TR_{fP}.$$

Note that the order of matrices is reversed because it is the final refracted matrix that transforms (acts on) the translation matrix, which in turn, transforms the first refraction matrix. Therefore,

$$S = R_{fQ}TR_{fP}.$$

Recall that

$$R_{fP} = \begin{pmatrix} 1 & \frac{n_1 - n_2}{r_1} \\ 0 & 1 \end{pmatrix}, \quad T = \begin{pmatrix} 1 & 0 \\ \frac{L}{n_2} & 1 \end{pmatrix}, \quad R_{fQ} = \begin{pmatrix} 1 & \frac{n_2 - n_3}{-r_2} \\ 0 & 1 \end{pmatrix}$$

We used $-r_2$, and not r_2 because the second surface is a concave surface.

$$\Rightarrow S = \begin{pmatrix} 1 & \frac{n_2 - n_3}{-r_2} \\ 0 & 1 \end{pmatrix} \begin{pmatrix} 1 & 0 \\ \frac{L}{n_2} & 1 \end{pmatrix} \begin{pmatrix} 1 & \frac{n_1 - n_2}{r_1} \\ 0 & 1 \end{pmatrix}$$

The refraction power P_1 of the first surface is

$$P_1 = \frac{n_2 - n_1}{r_1}$$

similarly,

$$P_2 = \frac{n_3 - n_2}{-r_2}$$

$$\Rightarrow S = \begin{pmatrix} 1 & -P_2 \\ 0 & 1 \end{pmatrix} \begin{pmatrix} 1 & 0 \\ \frac{L}{n_2} & 1 \end{pmatrix} \begin{pmatrix} 1 & -P_1 \\ 0 & 1 \end{pmatrix}$$

Note that

$$|S| = 1$$

$$S = \begin{pmatrix} 1 & -P_2 \\ 0 & 1 \end{pmatrix} \begin{pmatrix} 1 & -P_1 \\ \frac{L}{n_2} & 1 - \frac{P_1 L}{n_2} \end{pmatrix}$$

$$\Rightarrow S = \begin{pmatrix} 1 - \frac{P_2 L}{n_2} & -P_1 - P_2(1 - \frac{P_1 L}{n_2}) \\ \frac{L}{n_2} & 1 - \frac{P_1 L}{n_2} \end{pmatrix}$$

$$\Rightarrow S = \begin{pmatrix} 1 - \frac{P_2 L}{n_2} & \frac{P_1 P_2 L}{n_2} - P_1 - P_2 \\ \frac{L}{n_2} & 1 - \frac{P_1 L}{n_2} \end{pmatrix}$$

GAUSS CONSTANTS

In general, the system matrix S of a composite optical system is written as

$$S = \begin{pmatrix} B & -A \\ -D & C \end{pmatrix}$$

Where A, B, C, and D are called Gauss Constants. Comparing this general system matrix with that of thick lens matrix already established, we have

$$B = 1 - \frac{P_2 L}{n_2} \quad , \quad C = 1 - \frac{P_1 L}{n_2}$$

$$D = \frac{-L}{n_2} \quad , \quad A = P_1 + P_2 - \frac{P_1 P_2 L}{n_2}$$

Note that

$$|S| = BC - AD = 1$$

THIN LENS MATRIX

A lens is regarded as thin if the thickness L is such that $L << r_1, L << r_2$. In this case, we can approximate L as zero. That is, for a thin lens, $L \approx 0$.

$\Rightarrow B = 1$, $C = 1$, $D = 0$, $A = P_1 + P_2$. The system matrix becomes

$$S = \begin{pmatrix} 1 & -P_1 - P_2 \\ 0 & 1 \end{pmatrix} = \begin{pmatrix} 1 & -(P_1 + P_2) \\ 0 & 1 \end{pmatrix}$$

The total optical power of the thin lens becomes P given by

$$P = P_1 + P_2$$

$$\Rightarrow \qquad P = \frac{n_2 - n_1}{r_1} + \frac{n_3 - n_2}{-r_2}$$

$$\Rightarrow \qquad P = \frac{n_2 - n_1}{r_1} - \frac{n_3 - n_2}{r_2} \qquad\qquad ----- P178$$

If the medium on both sides of the lens is the same, $n_1 = n_3$.

$$\Rightarrow \qquad\qquad P = (n_2 - n_1)\left(\frac{1}{r_1} + \frac{1}{r_2}\right) \qquad\qquad ------ P179$$

$$\text{Optical power, } P = \frac{1}{focal \ lenght \ (f)}$$

$$\Rightarrow \frac{1}{f} = P = (n_2 - n_1)\left(\frac{1}{r_1} + \frac{1}{r_2}\right)$$

We should not forget that the system matrix acts on the input ray; that is, it transforms the input ray to give the output ray. In other words,

$$\begin{pmatrix} n_2\alpha_2 \\ y_2 \end{pmatrix} = S \begin{pmatrix} n_1\alpha_1 \\ y_1 \end{pmatrix}$$

The planes normal to the optic axis at the input and output points are respectively called the **input plane** and the **output plane.**

Example 4.1

Consider a thick hemispherical lens of glass ($n = 1.5$) having radius of curvature 5cm. Write the system matrix for the lens kept in air.

Solution

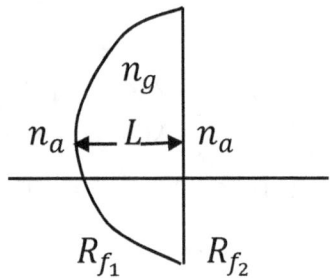

The system matrix is developed as follows: at the curved surface refraction occurs, then translation between the two surfaces, followed by refraction at the flat (straight line) surface.

$$R_{f_1} = \begin{pmatrix} 1 & \dfrac{n_g - n_a}{r_2} \\ 0 & 1 \end{pmatrix}, \quad T = \begin{pmatrix} 1 & 0 \\ \dfrac{L}{n_g} & 1 \end{pmatrix}, \quad R_{f_2} = \begin{pmatrix} 1 & \dfrac{n_a - n_g}{r_1} \\ 0 & 1 \end{pmatrix}$$

Where r_1 and r_2 are the radii of curvatures, respectively, of the two surfaces, n_a = air refractive index, n_g = glass refractive index, L = thickness of the glass. Note that the radius of curvature of a plane (flat) surface is infinite.

$$\Rightarrow \quad r_2 = \infty, \Rightarrow \frac{n_g - n_a}{r_2} = 0$$

$$\Rightarrow \quad S = \begin{pmatrix} 1 & 0 \\ 0 & 1 \end{pmatrix} \begin{pmatrix} 1 & 0 \\ \dfrac{L}{n_g} & 1 \end{pmatrix} \begin{pmatrix} 1 & \dfrac{n_a - n_g}{r_1} \\ 0 & 1 \end{pmatrix}$$

Note, also that

$$L = r_1; \; n_a = 1$$

$$\Rightarrow \quad S = \begin{pmatrix} 1 & 0 \\ \dfrac{5}{1.5} & 1 \end{pmatrix} \begin{pmatrix} 1 & \dfrac{1-1.5}{5} \\ 0 & 1 \end{pmatrix}$$

$$\Rightarrow S = \begin{pmatrix} 1 & 0 \\ \dfrac{10}{3} & 1 \end{pmatrix} \begin{pmatrix} 1 & \dfrac{-1}{10} \\ 0 & 1 \end{pmatrix}$$

$$\Rightarrow S = \begin{pmatrix} 1 & \dfrac{-1}{10} \\ \dfrac{10}{3} & \dfrac{2}{3} \end{pmatrix}$$

On comparing to the standard system matrix S given by

$$S = \begin{pmatrix} B & -A \\ -D & C \end{pmatrix}, \text{ we have the Gauss constants as}$$

$$B = 1, A = \frac{1}{10} \, cm^{-1}, D = \frac{-10}{3} \, cm, C = \frac{2}{3}$$

OBJECT- TO- IMAGE MATRIX

By now, we should be convinced that any composite optic system, no matter how complex, can be represented by a 2x2 matrix called the system matrix. This system matrix links the object to the image. Consider the diagram below as an illustration.

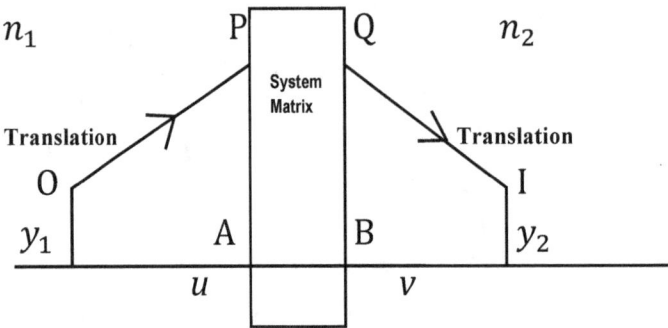

We take points A and B as the points of intersection of the input and output planes respectively with the optic axis.

In the diagram above, an object O is kept at a distance u to the left of the input plane, at height y_1 above optic axis. Let the image I of object O be formed at a distance v to the right of the output plane, at height y_2 above the optic axis. We wish to find relations between u and v, and y_1 and y_2. It should be noted that the image I is a point where all the rays coming from the object point O converge. As shown in the diagram, a ray OP is incident from O on the optical system. This ray then passes through the system, emerges at Q on the output plane and finally travels freely to the image point I.

From point O to point P, the ray undergoes translation T_1 in a medium of refractive index n_1; thereafter, it passes through the system, acted upon (transformed) by the system matrix. The output ray at Q then undergoes another translation T_2 from Q to I in a homogeneous medium of refractive index n_2. As before,

Let $\begin{pmatrix} n_1\alpha_1 \\ y_1 \end{pmatrix}$ be the coordinate of the ray at the initial (object) point O, and

$\begin{pmatrix} n_2\alpha_2 \\ y_2 \end{pmatrix}$, the ray coordinate at I. From our analysis,

$\begin{pmatrix} n_2\alpha_2 \\ y_2 \end{pmatrix} = T_2 S T_1 \begin{pmatrix} n_1\alpha_1 \\ y_1 \end{pmatrix}$, where $S = system\ matrix = \begin{pmatrix} B & -A \\ -D & C \end{pmatrix}$ and

$$T_2 = translation\ from\ Q\ to\ I = \begin{pmatrix} 1 & 0 \\ \dfrac{v}{n_2} & 1 \end{pmatrix},$$

$$T_1 = translation\ from\ O\ to\ P = \begin{pmatrix} 1 & 0 \\ \dfrac{-u}{n_1} & 1 \end{pmatrix}$$

Note: $-u$ because the distance is to the left of the optical system.

$$\Rightarrow \quad \begin{pmatrix} n_2\alpha_2 \\ y_2 \end{pmatrix} = \begin{pmatrix} 1 & 0 \\ \dfrac{v}{n_2} & 1 \end{pmatrix} \begin{pmatrix} B & -A \\ -D & C \end{pmatrix} \begin{pmatrix} 1 & 0 \\ \dfrac{-u}{n_1} & 1 \end{pmatrix} \begin{pmatrix} n_1\alpha_1 \\ y_1 \end{pmatrix}$$

$$\Rightarrow \begin{pmatrix} n_2\alpha_2 \\ y_2 \end{pmatrix} = \begin{pmatrix} 1 & 0 \\ \dfrac{v}{n_2} & 1 \end{pmatrix} \begin{pmatrix} B + \dfrac{Au}{n_1} & -A \\ -D - \dfrac{Cu}{n_1} & C \end{pmatrix} \begin{pmatrix} n_1\alpha_1 \\ y_1 \end{pmatrix}$$

$$\Rightarrow \begin{pmatrix} n_2\alpha_2 \\ y_2 \end{pmatrix} = \begin{pmatrix} B + \dfrac{Au}{n_1} & -A \\ \dfrac{v}{n_2}\left(B + \dfrac{Au}{n_1}\right) - D - \dfrac{Cu}{n_1} & \dfrac{-Av}{n_2} + C \end{pmatrix} \begin{pmatrix} n_1\alpha_1 \\ y_1 \end{pmatrix}$$

If we assume that the optical system is kept in air; that is, the media on both sides of the system is air, then $n_1 = n_2 = 1$

$$\Rightarrow \quad \begin{pmatrix} \alpha_2 \\ y_2 \end{pmatrix} = \begin{pmatrix} B + Au & -A \\ Bv + Avu - D - Cu & C - Av \end{pmatrix} \begin{pmatrix} \alpha_1 \\ y_1 \end{pmatrix}$$

$$\begin{pmatrix} \alpha_2 \\ y_2 \end{pmatrix} = M \begin{pmatrix} \alpha_1 \\ y_1 \end{pmatrix}$$

This 2 x 2 matrix M is called **object-to-image matrix**.

$$\Rightarrow \begin{pmatrix} \alpha_2 \\ y_2 \end{pmatrix} = \begin{pmatrix} (B + Au)\alpha_1 - Ay_1 \\ (Bv + Avu - D - Cu)\alpha_1 + (C - Av)y_1 \end{pmatrix}$$

$$\Rightarrow \quad \alpha_2 = (B + Au)\,\alpha_1 - Ay_1$$

$$y_2 = (Bv + Avu - D - Cu)\,\alpha_1 + (C - Av)y_1$$

Since any ray from O, irrespective of its angle with the optical axis, converges at I, this implies that y_2 is independent of α_1

$$\Rightarrow \quad \frac{dy_2}{d\alpha_1} = 0$$

$$\Rightarrow \quad Bv + Avu - D - Cu = 0 \qquad\qquad ----- P180$$

$$y_2 = (C - Av)\,y_1$$

The ratio $\frac{y_2}{y_1}$ is defined as the magnification m.

$$\Rightarrow \quad m = \frac{y_2}{y_1} = C - Av \qquad\qquad ----- P181$$

Using the fact that

$$|T_2 S T_1| = 1, we\ have$$

$$(C - Av)\,(B + Au) = 1$$

$$\Rightarrow B + Au = \frac{1}{C - Av} = \frac{1}{m} \qquad\qquad ----- P182$$

$$\Rightarrow \quad M = \begin{pmatrix} \dfrac{1}{m} & -A \\ 0 & m \end{pmatrix}$$

$$Where\ m = \frac{y_2}{y_1} = C - Av = \frac{1}{B + uA}$$

Thin – Lens Equation

Recall that the system matrix S, for the thin lens system is

$$S = \begin{pmatrix} 1 & -P_1 - P_2 \\ 0 & 1 \end{pmatrix}$$

$$B = C = 1, D = 0, A = P_1 + P_2 = \frac{1}{f}$$

$\Rightarrow m$, for thin lens, is

$$m = C - Av = 1 - \frac{v}{f} \qquad\qquad ----- P183$$

Or

$$\frac{1}{m} = B + Au$$

$$\Rightarrow \frac{1}{m} = 1 + \frac{u}{f} \qquad\qquad ----- P184$$

Again, since

$$Bv + Avu - D - Cu = 0$$

$$\Rightarrow \quad v + \frac{vu}{f} - u = 0$$

$$\Rightarrow \quad \frac{1}{v} - \frac{1}{u} = \frac{1}{f} \qquad\qquad ----- P185$$

Example 4.2

Consider a thick symmetrical biconvex lens with radius of curvature of spherical surfaces equal to 5cm, if the thickness of the lens at the centre is also 5cm and its refractive index is 1.5, find

a) Gauss constants for the lens

b) Position of image of a point object kept at a distance of 20cm from lens surface on its axis.

Solution

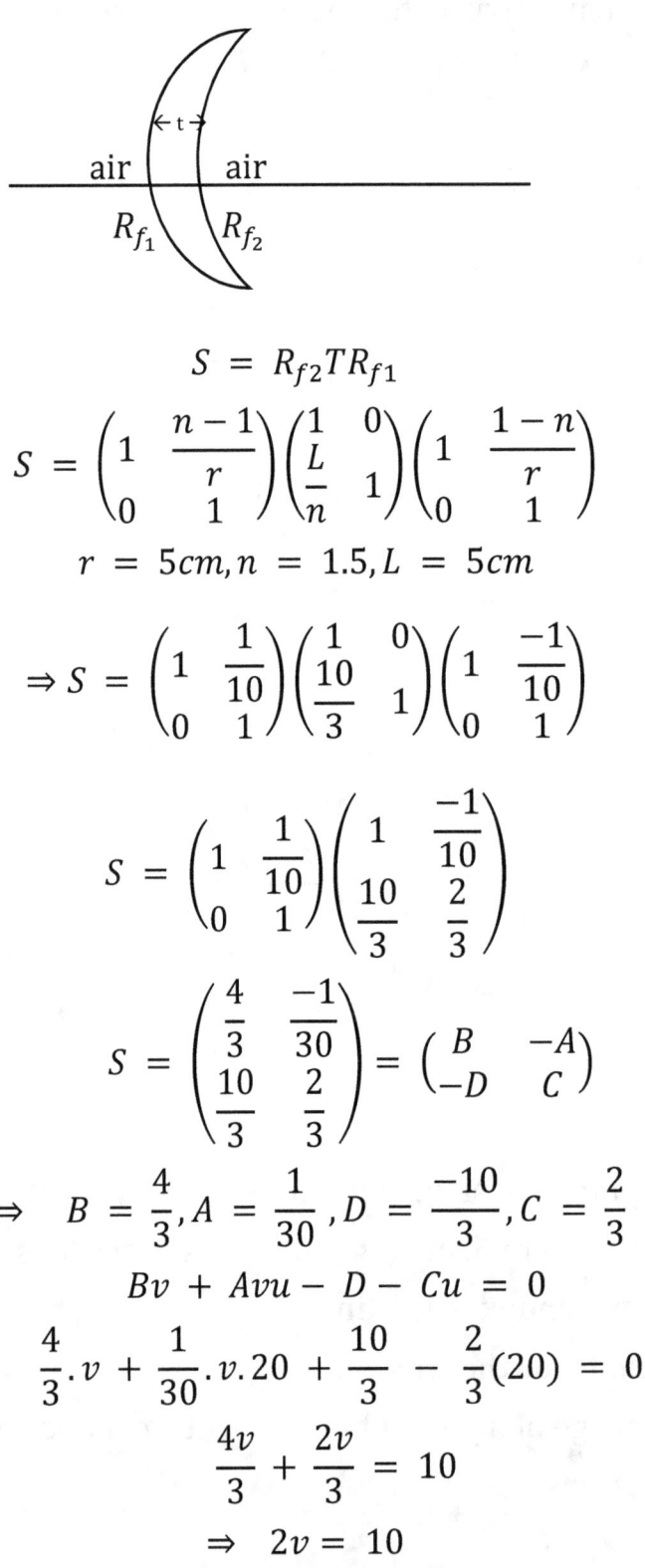

$$S = R_{f2}TR_{f1}$$

$$S = \begin{pmatrix} 1 & \dfrac{n-1}{r} \\ 0 & 1 \end{pmatrix}\begin{pmatrix} 1 & 0 \\ \dfrac{L}{n} & 1 \end{pmatrix}\begin{pmatrix} 1 & \dfrac{1-n}{r} \\ 0 & 1 \end{pmatrix}$$

$$r = 5cm, n = 1.5, L = 5cm$$

$$\Rightarrow S = \begin{pmatrix} 1 & \dfrac{1}{10} \\ 0 & 1 \end{pmatrix}\begin{pmatrix} 1 & 0 \\ \dfrac{10}{3} & 1 \end{pmatrix}\begin{pmatrix} 1 & \dfrac{-1}{10} \\ 0 & 1 \end{pmatrix}$$

$$S = \begin{pmatrix} 1 & \dfrac{1}{10} \\ 0 & 1 \end{pmatrix}\begin{pmatrix} 1 & \dfrac{-1}{10} \\ \dfrac{10}{3} & \dfrac{2}{3} \end{pmatrix}$$

$$S = \begin{pmatrix} \dfrac{4}{3} & \dfrac{-1}{30} \\ \dfrac{10}{3} & \dfrac{2}{3} \end{pmatrix} = \begin{pmatrix} B & -A \\ -D & C \end{pmatrix}$$

$$\Rightarrow \quad B = \frac{4}{3}, A = \frac{1}{30}, D = \frac{-10}{3}, C = \frac{2}{3}$$

$$Bv + Avu - D - Cu = 0$$

$$\Rightarrow \quad \frac{4}{3}.v + \frac{1}{30}.v.20 + \frac{10}{3} - \frac{2}{3}(20) = 0$$

$$\frac{4v}{3} + \frac{2v}{3} = 10$$

$$\Rightarrow \quad 2v = 10$$

$$\Rightarrow v = 5cm$$

Example 4.3

Consider a plano – convex glass lens (n = 1.5) of thickness 9cm and radius of curvature of convex surface equal to 2.5cm. Find

a) the element of the system matrix of the lens in air

b) where should an object be placed in front of the convex face of the lens so that the image is formed on its plane face?

Solution

$$S = \begin{pmatrix} 1 & \dfrac{1.5-1}{\infty} \\ 0 & 1 \end{pmatrix} \begin{pmatrix} 1 & 0 \\ \dfrac{9}{1.5} & 1 \end{pmatrix} \begin{pmatrix} 1 & \dfrac{1-1.5}{2.5} \\ 0 & 1 \end{pmatrix}$$

$$S = \begin{pmatrix} 1 & 0 \\ 0 & 1 \end{pmatrix} \begin{pmatrix} 1 & 0 \\ 6 & 1 \end{pmatrix} \begin{pmatrix} 1 & \dfrac{-1}{5} \\ 0 & 1 \end{pmatrix}$$

$$S = \begin{pmatrix} 1 & \dfrac{-1}{5} \\ 6 & \dfrac{-1}{5} \end{pmatrix} = \begin{pmatrix} B & -A \\ -D & C \end{pmatrix}$$

$$\Rightarrow B = 1, A = \frac{1}{5} cm^{-1}, D = -6cm, C = \frac{-1}{5}$$

b)
$$Bv + Avu - D - Cu = 0$$

Find u such that $v = 0$. This is the meaning of the second question.

$$\Rightarrow -D - Cu = 0$$

$$\Rightarrow u = \frac{-D}{C}$$

$$\Rightarrow u = \frac{--6}{\dfrac{-1}{5}}$$

$$\therefore u = -30cm$$

This implies that the object should be placed 30cm to the left of the convex surface.

Cardinal Points and Cardinal Planes of an Optical System

There are six points on the optic axis of an optical system called **cardinal points** which completely describe the properties of the system. The planes normal to the axis at these six points are called **cardinal planes.** The positions of the cardinal points, and hence the cardinal planes, are related to the Gauss constants (the entries of the system matrix) since the system matrix is repository of all the information necessary for a complete description of the optical system. These six cardinal points are defined as follows:

Principal Points

The first two cardinal points are called the principal points and their corresponding planes are called the principal planes- one in the object space, and the other in the image space labeled as shown below. The object space is the set of all points to the left of the optic centre. The principal planes are labeled G_1 and G_2 as shown below.

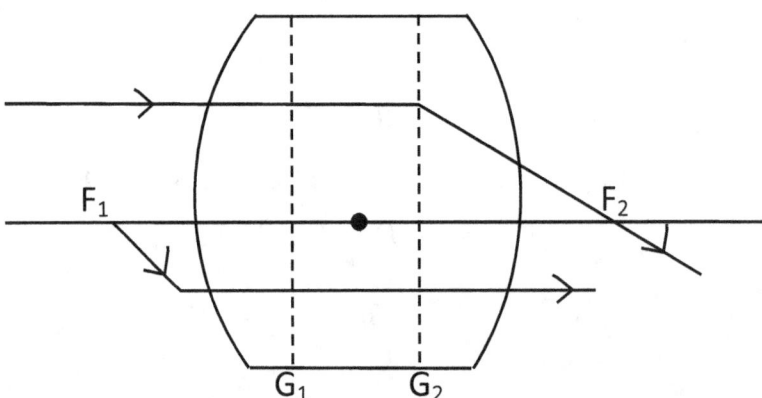

By definition, the principal plane G_1 is such that any paraxial ray starting from any height in G_1 finally passes through the same height in G_2 ; that is, if $\begin{pmatrix} n_1\alpha_1 \\ y_1 \end{pmatrix}$ is the ray coordinate on G_1, the ray coordinate on G_2 is $\begin{pmatrix} n_2\alpha_2 \\ y_2 = y_1 \end{pmatrix}$. Let $n_1 = n_2 = 1$.

$$\Rightarrow \quad m = \frac{y_2}{y_1} = 1$$

$$\Rightarrow C - vA = 1$$

$$\Rightarrow \qquad v = \frac{C-1}{A} = l_{G_2} \qquad ----- P186$$

Again,

$$\Rightarrow \quad \frac{1}{B+Au} = 1$$

$$\Rightarrow B + Au = 1$$

$$\Rightarrow u = \frac{1-B}{A} = l_{G_1} \qquad ----- P187$$

Where l_{G_1} and l_{G_2} are the distances of the principal planes G_1 and G_2 from the input and output planes of the optical system respectively.

For a thin lens, the two principal planes coincide; that is, they meet at the optic centre together with the input and out planes. Therefore

$$l_{G_1} = l_{G_2} = 0$$

$$\Rightarrow \frac{1-B}{A} = 0$$

$$\Rightarrow B = 1$$

$$\Rightarrow \frac{C-1}{A} = 0$$

$$\Rightarrow C = 1$$

Focal Points

Like the principal points, there are two focal points. The first focal point F_1, as shown in the diagram, is a point in the object space from where the incident rays after passing through the system become parallel to the optic axis. The second focal point, F_2, in the image space is the point at which incident rays parallel to the optic axis converge.

To determine F_1, we use its definition. The definition of F_1 implies that

$$y_1 = 0, \alpha_2 = 0$$

since

$$\alpha_2 = (B + Au)\,\alpha_1 - Ay_1$$
$$\Rightarrow 0 = (B + Au)\alpha_1$$

To avoid triviality, $\alpha_1 \neq 0$.

$$\Rightarrow B + Au = 0$$
$$\Rightarrow u = \frac{-B}{A} = l_{F_2}$$

Also, F_2 is such that $\alpha_1 = 0, y_2 = 0$

Since

$$y_2 = (Bv + Avu - D - Cu)\alpha_1 + (C - Av)y_1$$
$$\Rightarrow (C - Av)y_1 = 0$$

To avoid the trivial case, $y_1 \neq 0$

$$C - Av = 0$$
$$\Rightarrow v = \frac{C}{A} = l_{F_2}$$

Where l_{F_1} the distance of the first is focus to the left of the input plane, and l_{F_2} is the distance of the second focus to the right of the output plane.

Focal lengths

The distances of the focal planes from the respective principal planes are defined as the focal lengths of an optical system. Hence, the focal length in the object space is given by

$$f_1 = l_{G_1} - l_{F_1}$$
$$= \frac{-B}{A} - \left(\frac{1-B}{A}\right)$$

$$f_1 = \frac{-1}{A} \qquad\qquad ----- P188$$

Similarly, the second focal length (the focal length in the image space) is given by

$$f_2 = l_{G_2} - l_{F_2}$$

$$= \frac{C}{A} - \left(\frac{C-1}{A}\right)$$

$$f_2 = \frac{1}{A} \qquad ----- P189$$

Nodal Points

The two points N_1 and N_2 on the optic axis such that an incident ray directed towards the first point N_1 at angle α_1 finally emerges from the second point N_2 at the same angle $\alpha_2 = \alpha_1$. The planes passing through the nodal points and normal to the optic axis are called nodal planes.

To determine the position of the nodal point, we take a sample point $\begin{pmatrix} n_1\alpha_1 \\ y_1 = 0 \end{pmatrix}$ on N_1. This implies that the corresponding point on N_2 is $\begin{pmatrix} n_2\alpha_2 = n_2\alpha_1 \\ y_2 \end{pmatrix}$ since

$$\alpha_1 = \alpha_2.$$

Considering our equation in a more general form; that is, $n_1 \neq n_2$, we have

$$n_2\alpha_2 = (B + Au)\, n_1\alpha_1 - Ay_1$$

$$y_2 = (Bv + Avu - D - Cu)n_1\alpha_1 + (C - Av)y_1$$

Considering our sample point we have

$$n_2\alpha_2 = (B + Au)\, n_1\alpha_2 - 0$$

$$y_2 = (Bv + Avu - D - Cu)n_1\alpha_2$$

Since $Bv + Avu - D - Cu = 0$ (y_2 is independent of $\alpha_1 = \alpha_2$)

$$\Rightarrow \quad y_2 = 0$$

$$\Rightarrow \quad n_2\alpha_2 = (B + Au)n_1\alpha_2$$

$$\Rightarrow \quad \frac{n_2}{n_1} = B + Au$$

$$\Rightarrow \quad u = \frac{\frac{n_2}{n_1} - B}{A} = l_{N_1}$$

$$\Rightarrow \quad l_{N_1} = \frac{n_2 - Bn_1}{An_1}$$

When $n_1 = n_2$,

$$l_{N_1} = \frac{1 - B}{A}, \textit{which coincides with } l_{H1}$$

$$\Rightarrow \quad l_{N_1} = l_{H_1} \textit{ when } n_1 = n_2$$

Similarly,

$$l_{N_2} = l_{H_2} \textit{ when } n_1 = n_2$$

Cardinal points of a Thick Lens

Recall that for a thick lens

$$A = P_1 + P_2 - \frac{P_1 P_2 L}{n}$$

$$B = 1 - \frac{P_2 L}{n}$$

$$C = 1 - \frac{P_1 L}{n}$$

$$D = \frac{-L}{n}$$

Where L is the thickness of the optical system, n is the refractive index of the system, and P_1 and P_2 are the optical powers of the input and output surfaces respectively; from the facts we have established, it follows that for a thick lens,

$$l_{G_1} = \frac{1 - B}{A}$$

$$= \frac{1 - \left(1 - \frac{P_2 L}{n}\right)}{P_1 + P_2 - \frac{P_1 P_2 L}{n}}$$

$$\Rightarrow \qquad l_{G_1} = \frac{P_2 L}{nP_1 + nP_2 - P_1 P_2 L} \qquad ----- P190$$

$$l_{G_2} = \frac{C - 1}{A}$$

$$= \frac{1 - \dfrac{P_1 L}{n} - 1}{P_1 + P_2 - \dfrac{P_1 P_2 L}{n}}$$

$$\Rightarrow l_{G_2} = \frac{-P_1 L}{nP_1 + nP_2 - P_1 P_2 L} \qquad ----- P191$$

FOCAL LENGHTS OF A THICK LENS

Since

$\dfrac{1}{f} = A$, where f is the focal length,

$$\Rightarrow \frac{1}{f} = \frac{-P_1 P_2 L}{n} + P_1 + P_2$$

But

$$P_1 = \frac{n - 1}{r_1} \ , P_2 = \frac{1 - n}{r_2}$$

$$\Rightarrow \frac{1}{f} = \frac{-(n-1)(1-n)L}{nr_1 r_2} + \frac{n-1}{r_1} + \frac{1-n}{r_2}$$

$$\Rightarrow \frac{1}{f} = \frac{(n-1)^2}{nr_1 r_2} + (n-1)\left(\frac{1}{r_1} - \frac{1}{r_2}\right) \qquad ----- P192$$

Be reminded that f is the distance of the foci from the principal plane.

Thick Lens Equation

We define a new parameter U as the object-plane distance from the first principal plane. Similarly, V is the image-plane distance from the second principal plane. These imply that

$$u = l_{G_1} + U$$

$$\Rightarrow u = \frac{1 - B}{A} + U$$

Also,

$$v = l_{G_2} + V$$

$$\Rightarrow v = \frac{C - 1}{A} + V$$

$$\Rightarrow \quad Au + B = AU + 1$$

$$C - vA = -VA + 1$$

Since $|S| = 1$, where S is the system matrix,

$$\Rightarrow (Au + B)(C - vA) = 1$$

$$\Rightarrow (AU + 1)(1 - VA) = 1$$

$$\Rightarrow AU - A^2VU - VA + 1 = 1$$

$$\Rightarrow U - AVU - V = 0$$

$$\Rightarrow \quad U - V = AVU$$

On dividing through by UV, we have

$$\frac{1}{V} - \frac{1}{U} = A$$

$$\Rightarrow \frac{1}{V} - \frac{1}{U} = \frac{1}{f} \qquad ----- P193$$

System of Thin Lenses

We consider the optical system comprising of two thin lenses of focal length f_1 and f_2 kept at a distance L part as shown below.

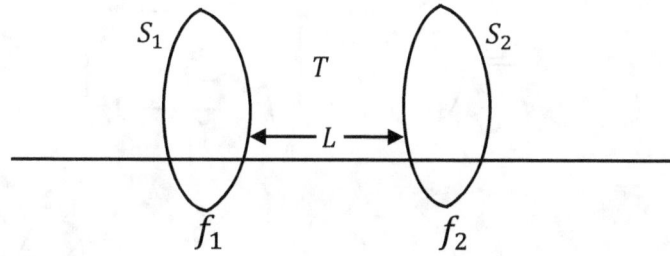

The overall system matrix is given by S.

$S = S_2 T S_1$ where S_1 and S_2 are the respective system matrices for the first and second optical systems (lenses) and T is the translation matrix between them.

Clearly,

$T = \begin{pmatrix} 1 & 0 \\ \frac{L}{n} & 1 \end{pmatrix}, n = 1$ because the whole system is assumed to be kept in air.

$$\Rightarrow \quad T = \begin{pmatrix} 1 & 0 \\ L & 1 \end{pmatrix}$$

$$S_1 = \begin{pmatrix} B_1 & -A_1 \\ -D_1 & C_1 \end{pmatrix}$$

$$B_1 = C_1 = 1, D_1 = 0$$

$$\Rightarrow S_1 = \begin{pmatrix} 1 & \dfrac{-1}{f_1} \\ 0 & 1 \end{pmatrix}$$

Similarly,

$$S_2 = \begin{pmatrix} 1 & \dfrac{-1}{f_2} \\ 0 & 1 \end{pmatrix}$$

$$S = S_2 T S_1$$

$$\Rightarrow S = \begin{pmatrix} 1 & \dfrac{-1}{f_2} \\ 0 & 1 \end{pmatrix} \begin{pmatrix} 1 & 0 \\ L & 1 \end{pmatrix} \begin{pmatrix} 1 & \dfrac{-1}{f_1} \\ 0 & 1 \end{pmatrix}$$

$$\Rightarrow \ S = \begin{pmatrix} 1 & \dfrac{-1}{f_2} \\ 0 & 1 \end{pmatrix} \begin{pmatrix} 1 & \dfrac{-1}{f_1} \\ L & \dfrac{-L}{f_1} + 1 \end{pmatrix}$$

$$\Rightarrow \ S = \begin{pmatrix} 1 - \dfrac{L}{f_2} & \dfrac{-1}{f_1} - \dfrac{1}{f_2}(1 - \dfrac{L}{f_1}) \\ L & 1 - \dfrac{L}{f_1} \end{pmatrix}$$

$$S = \begin{pmatrix} B & -A \\ -D & C \end{pmatrix}$$

$$\Rightarrow A = \frac{1}{f_2} + \frac{1}{f_2}\left(1 - \frac{L}{f_2}\right) \qquad\qquad ----P194$$

$f = \dfrac{1}{A}$, *the overall focal length of the whole system*

$$\Rightarrow \ \frac{1}{f} = \frac{1}{f_1} + \frac{1}{f_2}\left(1 - \frac{L}{f_1}\right) \qquad\qquad ------P195$$

$$B = 1 - \frac{L}{f_2} \qquad\qquad ------P196$$

$$C = 1 - \frac{L}{f_1} \qquad\qquad ------P197$$

$$D = -L$$

$$\Rightarrow \ l_{G_1} = \frac{1 - B}{A}$$

$$= \frac{1 - \left(1 - \dfrac{L}{f_2}\right)}{A}$$

$$= \frac{L}{Af_2}$$

$$= \frac{Lf}{f_2} \ ; \ since \ f = \frac{1}{A} \qquad\qquad ------P198$$

$$l_{G_2} = \frac{C-1}{A}$$

$$\Rightarrow \quad l_{G_2} = \frac{1 - \dfrac{L}{f_1} - 1}{A}$$

$$\Rightarrow l_{G_2} = \frac{-L}{Af_1}$$

$$\Rightarrow l_{G_2} = \frac{-Lf}{f_1} \qquad\qquad ----- P199$$

If $L = 0$; that is, for the system

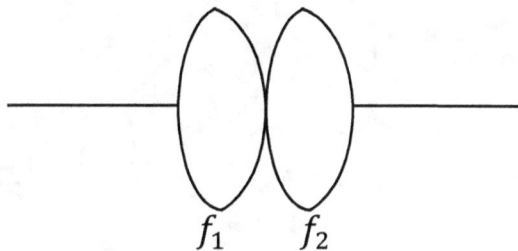

$$f_1 \qquad f_2$$

The overall focal length f is such that

$$\frac{1}{f} = \frac{1}{f_1} + \frac{1}{f_2}\left(1 - \frac{0}{f_1}\right)$$

$$\Rightarrow \frac{1}{f} = \frac{1}{f_1} + \frac{1}{f_2} \qquad\qquad ----- P200$$

This implies that for the system

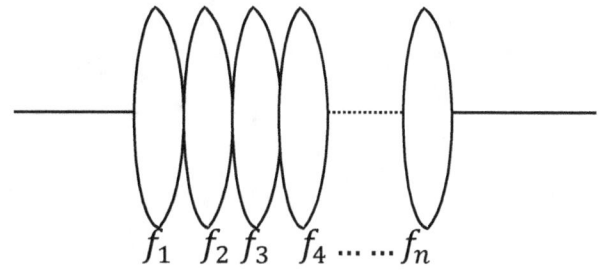

$$f_1 \quad f_2\, f_3 \quad f_4 \cdots \cdots f_n$$

$$\frac{1}{f} = \frac{1}{f_1} + \frac{1}{f_2} + \frac{1}{f_3} + \dots\dots + \frac{1}{f_n} \qquad ----- P201$$

For the system

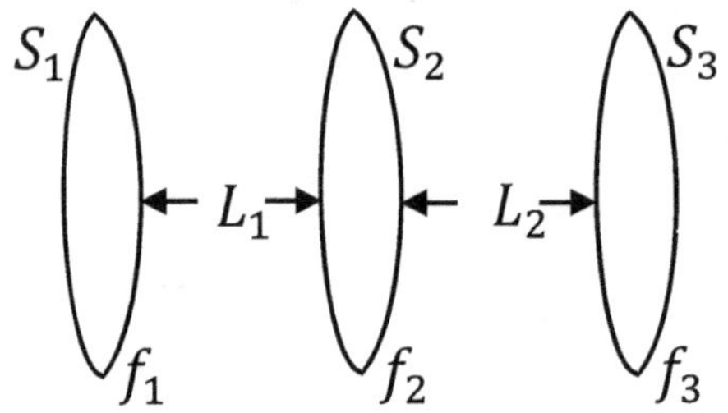

Lef f be the overall focal length for S_1 and S_2

$$\Rightarrow \frac{1}{f} = \frac{1}{f_1} + \frac{1}{f_2}(1 - \frac{L_1}{f_1})$$

For the whole system, let f_0 be the overall focal length.

$\Rightarrow \frac{1}{f_0} = \frac{1}{f_3} + \frac{1}{f}\left(1 - \frac{L_2}{f_3}\right)$, since L_2 is the distance of S_3 from the composite system S_1S_2. Therefore,

$$\frac{1}{f_0} = \frac{1}{f_3} + \left[\frac{1}{f_1} + \frac{1}{f_2}\left(1 - \frac{L_1}{f_1}\right)\right]\left(1 - \frac{L_2}{f_3}\right) \qquad ------P202$$

Chapter Five

Eigenvalue and Eigenvector:Eigenspace

Let A be an $n x n$ matrix. A scalar λ is called an **eigenvalue** of the matrix A corresponding to the nonzero eigenvector x if

$$Ax = \lambda x \qquad\qquad ----- P203$$

To find λ, we multiply through by I

$$IAx = I\lambda x$$
$$\Rightarrow Ax - I\lambda x = 0$$
$$\Rightarrow (A - \lambda I)x = 0$$

$$(\lambda I - A)x = 0 \qquad\qquad ----- P204$$

Equation P is the matrix equation for a homogeneous linear system having $(\lambda I - A)$ as its coefficient matrix. We already know that for this system to admit nontrivial solution vectors x, we must have

$$|\lambda I - A| = 0 \qquad\qquad ----- P205$$

Equation $P205$ is called **characteristic equation.** Solving equation $P205$ yields the *eigenvalues* of the matrix A.

The polynomial $|\lambda I - A|$ is called the **characteristic polynomial.**

Example 5.1

Find all the eigenvalues, if any, of the matrix

$$A = \begin{pmatrix} 1 & 0 & 1 \\ 1 & 2 & 0 \\ 3 & 0 & 1 \end{pmatrix}$$

Solution

A Matrix B, being a scalar matrix, may not have eigenvalues because the roots of the polynomial equation $|\lambda I - B| = 0$ may all be complex.

If B is viewed (considered) as a complex matrix, it will always have eigenvalues.

In this case, however, matrix A, whether considered real or complex, will always have eigenvalues simply because a cubic polynomial always has a real root (at least one).

If λ is an eigenvalue of A, then

$$|\lambda I - A| = 0$$

$$\Rightarrow \left| \begin{pmatrix} \lambda & 0 & 0 \\ 0 & \lambda & 0 \\ 0 & 0 & \lambda \end{pmatrix} - \begin{pmatrix} 1 & 0 & 1 \\ 1 & 2 & 0 \\ 3 & 0 & 1 \end{pmatrix} \right| = 0$$

$$\Rightarrow \begin{vmatrix} \lambda - 1 & 0 & -1 \\ -1 & \lambda - 2 & 0 \\ -3 & 0 & \lambda - 1 \end{vmatrix} = 0$$

Expanding along the second column, we have

$$(-1)^4 \, (\lambda - 2) \left((\lambda - 1)(\lambda - 1) - (-3)(-1) \right) = 0$$

$$\Rightarrow \quad (\lambda - 2) \left(((\lambda - 1)^2 - 3 \right) = 0$$

$$\Rightarrow \quad (\lambda - 2) = 0 \; or \; (\lambda - 1)^2 - 3 = 0$$

$$\Rightarrow \; \lambda = 2 \; or \; \lambda = 1 \pm \sqrt{3}$$

Therefore the eigenvalues are $\lambda = 2, \lambda = 1 + \sqrt{3}, \lambda = 1 - \sqrt{3}$

To find the respective eigenvectors, we need to solve the homogeneous linear system

$(\lambda I - A)x = 0$ for each λ. For $\lambda = 2$, we have

$$(2I - A)x = 0$$

$$\Rightarrow \begin{pmatrix} 1 & 0 & -1 \\ -1 & 0 & 0 \\ -3 & 0 & 1 \end{pmatrix} \begin{pmatrix} x_1 \\ x_2 \\ x_3 \end{pmatrix} = \begin{pmatrix} 0 \\ 0 \\ 0 \end{pmatrix}$$

$\Rightarrow \quad x_1 = 0, x_3 = x_1 = 0, x_2 = s$; where s is any real number.

$$\Rightarrow \quad \begin{pmatrix} x_1 \\ x_2 \\ x_3 \end{pmatrix} = \begin{pmatrix} 0 \\ s \\ 0 \end{pmatrix} = s \begin{pmatrix} 0 \\ 1 \\ 0 \end{pmatrix}$$

The eigenvector corresponding to the eigenvalue $\lambda = 2$ is

$$x = \begin{pmatrix} 0 \\ 1 \\ 0 \end{pmatrix}$$

The linear span of this vector is the eigenspace of A corresponding to the eigenvalue $\lambda = 2$. This eigenspace is of dimension one because the number of linearly independent vectors that span it is one.

For $\lambda = 1 + \sqrt{3}$, we have

$$((1 + \sqrt{3})I - A)x = 0$$

$$\Rightarrow \begin{pmatrix} \sqrt{3} & 0 & -1 \\ -1 & -1 + \sqrt{3} & 0 \\ -3 & 0 & \sqrt{3} \end{pmatrix} \begin{pmatrix} x_1 \\ x_2 \\ x_3 \end{pmatrix} = \begin{pmatrix} 0 \\ 0 \\ 0 \end{pmatrix}$$

$$\begin{pmatrix} \sqrt{3} & 0 & -1 & | & 0 \\ -1 & -1+\sqrt{3} & 0 & | & 0 \\ -3 & 0 & \sqrt{3} & | & 0 \end{pmatrix} (R_3 + R_1\sqrt{3}) \begin{pmatrix} \sqrt{3} & 0 & -1 & | & 0 \\ -1 & -1+\sqrt{3} & 0 & | & 0 \\ 0 & 0 & 0 & | & 0 \end{pmatrix}$$

$$(R_1 + \sqrt{3}\,R_2), (R_1 \sim R_2) \begin{pmatrix} -1 & -1+\sqrt{3} & 0 & | & 0 \\ 0 & 3-\sqrt{3} & -1 & | & 0 \\ 0 & 0 & 0 & | & 0 \end{pmatrix}$$

$$\Rightarrow x_3 = (3 - \sqrt{3})x_2, x_1 = (1 + \sqrt{3})x_2$$

$$\Rightarrow x = \begin{pmatrix} x_1 \\ x_2 \\ x_3 \end{pmatrix} = \begin{pmatrix} (1+\sqrt{3})x_2 \\ x_2 \\ (3-\sqrt{3})x_2 \end{pmatrix} = x_2 \begin{pmatrix} 1+\sqrt{3} \\ 1 \\ 3-\sqrt{3} \end{pmatrix}$$

Therefore, the eigenvector corresponding to $\lambda = 1 + \sqrt{3}$ is

$x = \begin{pmatrix} 1 + \sqrt{3} \\ 1 \\ 3 - \sqrt{3} \end{pmatrix}$. It is of dimension one because its basis is the singleton set

$$G = \{(1 + \sqrt{3}, 1, 3 - \sqrt{3})\}.$$

For $\lambda = 1 - \sqrt{3}$, we have

$$|(1 - \sqrt{3})I - A| = 0$$

$$\Rightarrow \begin{pmatrix} -\sqrt{3} & 0 & -1 \\ -1 & -1 - \sqrt{3} & 0 \\ -3 & 0 & -\sqrt{3} \end{pmatrix} \begin{pmatrix} x_1 \\ x_2 \\ x_3 \end{pmatrix} = \begin{pmatrix} 0 \\ 0 \\ 0 \end{pmatrix}$$

$$\Rightarrow -3x_1 = \sqrt{3}x_3, \qquad -x_1 = (1 + \sqrt{3})x_2, \qquad x_3 = -\sqrt{3}x_1$$

$$\Rightarrow x_3 = -\sqrt{3}x_1, \quad x_2 = \frac{-1}{1 + \sqrt{3}} x_1$$

$$\Rightarrow x_3 = -\sqrt{3}x_1, \quad x_2 = \frac{-1 + \sqrt{3}}{2} x_1$$

$$\Rightarrow x = \begin{pmatrix} x_1 \\ x_2 \\ x_3 \end{pmatrix} \begin{pmatrix} x_1 \\ \dfrac{1 + \sqrt{3}}{2} x_1 \\ -\sqrt{3}\, x_1 \end{pmatrix} = \tfrac{1}{2} x_1 \begin{pmatrix} 2 \\ 1 + \sqrt{3} \\ -2\sqrt{3} \end{pmatrix}$$

Therefore, the eigenvector corresponding to the eigenvalue $\lambda = 1 - \sqrt{3}$ is

$x = \begin{pmatrix} 2 \\ 1 + \sqrt{3} \\ 1 - 2\sqrt{3} \end{pmatrix}$. The corresponding eigenspace is spanned by this vector, and is

of dimension one.

Example 5.2

Find the eigenvector corresponding to each eigenvalue of the following matrices.

a) $A = \begin{pmatrix} 3 & 1 \\ 1 & 3 \end{pmatrix}$ (b) $A = \begin{pmatrix} 1 & 2 & 0 \\ 0 & 2 & -1 \\ 0 & 0 & 3 \end{pmatrix}$

Solution

a) If λ is an eigenvalue of A, then it satisfies the quadratic polynomial equation

$$|\lambda I - A| = 0$$

$$\Rightarrow \begin{vmatrix} \lambda - 3 & -1 \\ -1 & \lambda - 3 \end{vmatrix} = 0$$

$$\Rightarrow (\lambda - 3)2 - 1 = 0$$

$$\Rightarrow \lambda = 3 \pm 1$$

$$\therefore \lambda = 4 \; or \; 2$$

To find the eigenvector corresponding to $\lambda = 4$, we solve the linear homogeneous system

$$|4I - A| = 0$$

$$\Rightarrow \begin{pmatrix} 4 - 3 & -1 \\ -1 & 4 - 3 \end{pmatrix} \begin{pmatrix} x_1 \\ x_2 \end{pmatrix} \begin{pmatrix} 0 \\ 0 \end{pmatrix}$$

$$\Rightarrow \begin{pmatrix} 1 & -1 \\ -1 & 1 \end{pmatrix} \begin{pmatrix} x_1 \\ x_2 \end{pmatrix} \begin{pmatrix} 0 \\ 0 \end{pmatrix}$$

$$\Rightarrow x_1 = x_2$$

$x = \begin{pmatrix} x_1 \\ x_2 \end{pmatrix} = \begin{pmatrix} x_1 \\ x_1 \end{pmatrix} = x_1 \begin{pmatrix} 1 \\ 1 \end{pmatrix}$; where x_1 is any real number. Therefore the

eigenvector corresponding to the eigenvalue $\lambda = 4$ is $x = \begin{pmatrix} 1 \\ 1 \end{pmatrix}$.

This also means that the basis for the eigenspace of A corresponding to the eigenvalue $\lambda = 4$ is $G = \{(1,1)\}$.

For $\lambda = 2$, we have

$$\begin{pmatrix} -1 & -1 \\ -1 & -1 \end{pmatrix} \begin{pmatrix} x_1 \\ x_2 \end{pmatrix} = \begin{pmatrix} 0 \\ 0 \end{pmatrix}$$

$$\Rightarrow x_1 = -x_2 = s; \text{ where } s \text{ is any real number.}$$

$$\therefore x = \begin{pmatrix} x_1 \\ x_2 \end{pmatrix} = \begin{pmatrix} s \\ -s \end{pmatrix} or \begin{pmatrix} -s \\ s \end{pmatrix}$$

$$\Rightarrow \quad x = s\begin{pmatrix} 1 \\ -1 \end{pmatrix} \text{ or } s\begin{pmatrix} -1 \\ 1 \end{pmatrix}$$

Therefore the eigenvector of A corresponding to $\lambda = 2$ is

$x = \begin{pmatrix} 1 \\ -1 \end{pmatrix}$ or $x = \begin{pmatrix} -1 \\ 1 \end{pmatrix}$ – either of the two, but not both of them at a time.

Again, this eigenspace is of dimension one.

b) If λ is an eigenvalue of A, then it satisfies the cubic polynomial

$$|\lambda I - A| = 0$$

$$\Rightarrow \begin{vmatrix} \lambda - 1 & -2 & 0 \\ 0 & \lambda - 2 & 1 \\ 0 & 0 & \lambda - 3 \end{vmatrix} = 0$$

$$\Rightarrow \quad (\lambda - 1)(\lambda - 2)(\lambda - 3) = 0$$

$$\Rightarrow \quad \lambda = 1 \text{ or } 2 \text{ or } 3$$

Recall that the determinant of an upper triangular or lower triangular matrix is simply the product of elements along the main diagonal.

For $\lambda = 1$, we have

$$\begin{pmatrix} 0 & -2 & 0 \\ 0 & -1 & 1 \\ 0 & 0 & -2 \end{pmatrix}\begin{pmatrix} x_1 \\ x_2 \\ x_3 \end{pmatrix} = \begin{pmatrix} 0 \\ 0 \\ 0 \end{pmatrix}$$

$$\Rightarrow \quad x_1 = s \quad , \quad x_2 = -x_3 \quad , \quad 2x_2 = 0$$

$$\Rightarrow \begin{pmatrix} x_1 \\ x_2 \\ x_3 \end{pmatrix} = \begin{pmatrix} s \\ 0 \\ 0 \end{pmatrix} = s\begin{pmatrix} 1 \\ 0 \\ 0 \end{pmatrix} ; \text{ where } s \text{ is any real number.}$$

Therefore the eigenvector corresponding to $\lambda = 1$ is $x = \begin{pmatrix} 1 \\ 0 \\ 0 \end{pmatrix}$.

For $\lambda = 2$, we have

$$\begin{pmatrix} 1 & -2 & 0 \\ 0 & 0 & 1 \\ 0 & 0 & -1 \end{pmatrix}\begin{pmatrix} x_1 \\ x_2 \\ x_3 \end{pmatrix} = \begin{pmatrix} 0 \\ 0 \\ 0 \end{pmatrix}$$

$$\Rightarrow x_3 = 0, x_1 = 2x_2$$

$$\Rightarrow \quad x = \begin{pmatrix} x_1 \\ x_2 \\ x_3 \end{pmatrix} = \begin{pmatrix} 2x_2 \\ x_2 \\ 0 \end{pmatrix} = x_2\begin{pmatrix} 2 \\ 1 \\ 0 \end{pmatrix} ; \ x_2 \text{ is any real number.}$$

The eigenvector for $\lambda = 2$ is $x = \begin{pmatrix} 2 \\ 1 \\ 0 \end{pmatrix}$.

For $\lambda = 3$, we have

$$\begin{pmatrix} 2 & 2 & 0 \\ 0 & 1 & -1 \\ 0 & 0 & 0 \end{pmatrix} \begin{pmatrix} x_1 \\ x_2 \\ x_3 \end{pmatrix} = \begin{pmatrix} 0 \\ 0 \\ 0 \end{pmatrix}$$

$$\Rightarrow x_2 = x_3, 2x_1 = -2x_2$$

$$\Rightarrow x_2 = x_3, x_1 = -x_2$$

$$\Rightarrow x = \begin{pmatrix} x_1 \\ x_2 \\ x_3 \end{pmatrix} = \begin{pmatrix} -x_2 \\ x_2 \\ x_2 \end{pmatrix} = x_2 \begin{pmatrix} -1 \\ 1 \\ 1 \end{pmatrix}$$

Therefore, the eigenvector corresponding to $\lambda = 3$ is $x = \begin{pmatrix} -1 \\ 1 \\ 1 \end{pmatrix}$. The corresponding eigenspace is of dimension one.

Example 5.3

Determine the eigenvalues of $A = \begin{pmatrix} -1 & 0 & 0 \\ 0 & 3 & 1 \\ 0 & 1 & 3 \end{pmatrix}$

Solution

Let λ be an eigenvalue of A, then

$$|\lambda I - A| = 0$$

$$\Rightarrow \begin{vmatrix} \lambda + 1 & 0 & 0 \\ 0 & \lambda - 3 & 1 \\ 0 & -1 & \lambda - 3 \end{vmatrix} = 0$$

Expanding along the first row or first column, we have

$$(\lambda + 1)\left((\lambda - 3)^2 - 1\right) = 0$$

$$\Rightarrow \lambda = -1 \text{ or } (\lambda - 3)^2 = 1$$

$$\Rightarrow \lambda = -1 \text{ or } \lambda = 3 \pm 1$$

$$\Rightarrow \lambda = -1, 4, 2.$$

Example 5.4

Determine the basis of the eigenspace corresponding to each of the eigenvalues of

$$A = \begin{pmatrix} 4 & 2 & 0 \\ 2 & 4 & 0 \\ 0 & 0 & 2 \end{pmatrix}$$

Solution

Let λ be an eigenvalue of A, then

$$|\lambda I - A| = 0$$

$$\Rightarrow \begin{vmatrix} \lambda - 4 & -2 & 0 \\ -2 & \lambda - 4 & 0 \\ 0 & 0 & \lambda - 2 \end{vmatrix} = 0$$

Expanding along the third row or column, we have

$$(\lambda - 2)\left((\lambda - 4)^2 - 4\right) = 0$$

$$\Rightarrow (\lambda - 2)(\lambda - 4 - 2)(\lambda - 4 + 2) = 0$$

$$\Rightarrow \lambda = 2, \text{repeated}, \ \lambda = 6$$

For $\lambda = 2$, we have

$$\begin{pmatrix} -2 & -2 & 0 \\ -2 & -2 & 0 \\ 0 & 0 & 0 \end{pmatrix} \begin{pmatrix} x_1 \\ x_2 \\ x_3 \end{pmatrix} = \begin{pmatrix} 0 \\ 0 \\ 0 \end{pmatrix}$$

$$\Rightarrow -2x_1 = 2x_2 \ , \ x_3 = t \ , \ t \text{ is any real number.}$$

$$\Rightarrow x_1 = -x_2 = s \ , \ x_3 = t; \ s, t \text{ are any real numbers.}$$

$$\Rightarrow x = \begin{pmatrix} x_1 \\ x_2 \\ x_3 \end{pmatrix} = \begin{pmatrix} s \\ -s \\ t \end{pmatrix} = s\begin{pmatrix} 1 \\ -1 \\ 0 \end{pmatrix} + t\begin{pmatrix} 0 \\ 0 \\ 1 \end{pmatrix}$$

This implies that the eigenvalue $\lambda = 2$ has two corresponding eigenvectors

$\begin{pmatrix} 1 \\ -1 \\ 0 \end{pmatrix}$ and $\begin{pmatrix} 0 \\ 0 \\ 1 \end{pmatrix}$. This also means tha the eigenspace of A corresponding to $\lambda = 2$

is spanned by the two eigenvectors. However, we need to establish the linear independence of these vectors to conclude that the set $G = \{(1,-1,0),(0,0,1)\}$ forms a basis for the eigenspace of A corresponding to $\lambda = 2$.

$$\begin{pmatrix} 1 & -1 & 0 \\ 0 & 0 & 1 \end{pmatrix}$$

It is obvious that these two eigenvectors already form a row-echelon matrix having no zero row. Therefore, the two eigenvectors are linearly independent, and thus they form a basis for the eigenspace of A corresponding to eigenvalue $\lambda = 2$. This eigenspace is of dimension two, and its basis is the set $G = \{(1,-1,0),(0,0,1)\}$. This last result is not an accident.

It is always the case that if A is an nxn symmetric matrix, and λ is an eigenvalue of A with multiplicity $k \leq n$, then the eigenspace of A corresponding to λ must be spanned by k linearly independent eigenvectors – meaning that the corresponding eigenspace must be of dimension k.

This is not generally true for non-symmetric matrices although certain non symmetric matrices also exhibit this property.

For $\lambda = 6$, we have

$$\begin{pmatrix} 2 & -2 & 0 \\ -2 & 2 & 0 \\ 0 & 0 & 4 \end{pmatrix}\begin{pmatrix} x_1 \\ x_2 \\ x_3 \end{pmatrix} = \begin{pmatrix} 0 \\ 0 \\ 0 \end{pmatrix}$$

$$\Rightarrow \quad 4x_3 = 0, 2x_1 = 2x_2$$

$$\Rightarrow \quad x = \begin{pmatrix} x_1 \\ x_2 \\ x_3 \end{pmatrix} = \begin{pmatrix} x_1 \\ x_1 \\ 0 \end{pmatrix} = x_1 = \begin{pmatrix} 1 \\ 1 \\ 0 \end{pmatrix}$$

Therefore, the eigenvector of A corresponding to $\lambda = 6$ is

$x = \begin{pmatrix} 1 \\ 1 \\ 0 \end{pmatrix}$. Therefore, the basis for the eigenspace of A corresponding to $\lambda = 6$ is

of dimension one.

It is not difficult to see that if λ is an eigenvalue of A corresponding to the eigenvector $\boldsymbol{x}$, then

$$(\lambda I - A)\boldsymbol{x} = \boldsymbol{0}$$

implies that the matrix $(\lambda I - A)$ is a linear transformation on R^n (if A is an $n \times n$ matrix) whose kernel is the set $\{\boldsymbol{x} = (x_1, x_2, \dots x_n) | x_1, x_2 \dots \dots x_n \in R\}$. In other words, the kernel of the linear transformation whose matrix representation is $(\lambda I - A)$ is the eigenspace of A corresponding to the eigenvalue λ of A.

This further implies that determining the eigenspace of a matrix is equivalent to determining the kernel of some linear transformation. In other words, finding the eigenspace of A corresponding to λ is equivalent to finding the kernel of the linear transformation $(\lambda I - A)$. This is so because all linear transformations in a finite dimensional vector space can be represented as matrices.

Let λ be an eigenvalue of a real matrix A, then the only eigenvalue of the matrix $(\lambda I - A)$ is zero.

Proof

If λ is an eigenvalue of A with eigenvector $\boldsymbol{x}$, then

$$A\boldsymbol{x} = \lambda \boldsymbol{x}$$
$$\Rightarrow \quad \lambda I \boldsymbol{x} - A\boldsymbol{x} = \lambda I \boldsymbol{x} - \lambda \boldsymbol{x}$$
$$\Rightarrow (\lambda I - A)\boldsymbol{x} = \lambda \boldsymbol{x} - \lambda \boldsymbol{x}$$
$$\Rightarrow (\lambda I - A)\boldsymbol{x} = \boldsymbol{0}$$
$$\Rightarrow \quad (\lambda I - A)\boldsymbol{x} = k\boldsymbol{x} = \boldsymbol{0}$$
$$\Rightarrow \quad k = 0 \text{ since } \boldsymbol{x} \neq \boldsymbol{0} \text{ being an eigenvector of } A.$$

In general, to compute the eigenvalues and the corresponding eigenvectors of some linear transformation $T : V \rightarrow V$, one simply needs to select the matrix representation A of T relative to any basis for V, and then solve the characteristic equation for A. The eigenvalues for A are those for T. It should be noted that the eigenvalues for a linear transformation T is independent of the basis used in representing T as a matrix. In other words, the eigenvalues of a linear transformation T is basis invariant.

Let $T:V \rightarrow V$ be a linear transformation on the finite dimensional vector space V. If S and S' are bases for V, and if A and A' are the respective matrix representations for T relative to S and S', then the set of eigenvalues of A and A' are equal. The converse is generally not true; that is, the fact that two matrices have the same eigenvalues does not mean that the matrices are similar.

Proof

Since A and A' are matrix representations of T, they are similar.

This implies that there is a nonsingular matrix P such that

$$A = PA'P^{-1}$$
$$\Rightarrow \lambda I - A = \lambda I - PA'P^{-1}$$

But

$$PP^{-1} = I$$
$$\Rightarrow \lambda I - A = \lambda PP^{-1} - PA'P^{-1}$$
$$\Rightarrow \lambda I - A = P\lambda P^{-1} - PA'P^{-1}$$
$$\lambda I - A = P(\lambda I - A')P^{-1}$$
$$\Rightarrow |\lambda I - A| = |P(\lambda I - A')P^{-1}|$$
$$\Rightarrow |\lambda I - A| = |P||P^{-1}||\lambda I - A'| \text{, a property of determinant.}$$
$$\Rightarrow |\lambda I - A| = |\lambda I - A'| \text{; } since \ |P||P^{-1}| = 1$$

This means that any λ that satisfies the equation $|\lambda I - A| = 0$ also satisfies the equation $|\lambda I - A'| = 0$. Therefore, A and A' have the same eigenvalues.

Moreover, if x is an eigenvector for A, then $P^{-1}x$ is an eigenvector for A' because

$$Ax = \lambda x$$
$$\Rightarrow P^{-1}Ax = P^{-1}\lambda x$$

But

$$A = PA'P^{-1}$$

$$\Rightarrow P^{-1}A = A'P^{-1} \text{, upon left multiplying both sides by } P^{-1}$$

$$\Rightarrow A'P^{-1}x = P^{-1}\lambda x$$

$$\Rightarrow A'(P^{-1}x) = \lambda(P^{-1}x)$$

$$\Rightarrow \quad P^{-1}x \text{ is an eigenvector of } A' \text{ corresponding to the eigenvalue } \lambda.$$

Example 5.5

Let T be the transformation in the space P_2 of quadratic polynomials given by

$$T(c + bx + ax^2) = (4c + a) + (b - 2c)x + (a - 2c)x^2$$

a) Write the matrix representation for T relative to the bases

$E = \{1, x, x^2\}, and \ H = \{1, 1 - x, 1 - x^2\}$ and find the eigenvalues of T.

b) Find the bases for the corresponding eigenspaces of T relative to G and H.

Solution

$$E = \{1, x, x^2\} \equiv \{(1,0,0), (0,1,0), (0,0,1)\}$$

$$T(1,0,0) = 4 + (0 - 2)x + (0 - 2)x^2$$

$$\equiv (4, -2, -2)$$

$$T(0,1,0) = 0 + (1 - 2(0))x + 0. x^2$$

$$\equiv (0,1,0)$$

$$T(0,0,1) = 1 + x^2$$

$$\equiv (1, 0, 1)$$

Let the matrix representation for T relative to E be $[T]_E$.

$$[T]_E = \begin{pmatrix} 4 & 0 & 1 \\ -2 & 1 & 0 \\ -2 & 0 & 1 \end{pmatrix}$$

If λ is an eigenvalue of $[T]_E$, then

$$|\lambda I - [T]_E| = 0$$

$$\Rightarrow \begin{vmatrix} \lambda - 4 & 0 & -1 \\ 2 & \lambda - 1 & 0 \\ 2 & 0 & \lambda - 1 \end{vmatrix} = 0$$

Expanding along the first row, we have

$$\Rightarrow (\lambda - 4)\,((\lambda - 1)^2 - 0) + 2(\lambda - 1) = 0$$
$$\Rightarrow (\lambda - 1)\,((\lambda - 4)(\lambda - 1) + 2) = 0$$
$$\Rightarrow (\lambda - 1)(\lambda^2 - 5\lambda + 6) = 0$$
$$\Rightarrow \lambda = 1, 2, 3$$

Now we represent T relative to the basis H.

$$H = \{1, 1 - x, 1 - x^2\} \equiv \{(1,0,0), (1,-1,0), (1,0,-1)\}$$
$$T(1,0,0) \equiv (4,-2,-2)$$
$$T(1,-1,0) \equiv (4,-3,-2)$$
$$T(1,0,-1) \equiv (3,-2,-3)$$

The next thing is to $H - coordinatize$ each of $T(1,0,0)$, $T(1,-1,0)$, and $T(1,0,-1)$, after which we arrange them in column to form the matrix representation for T relative to H.

$$[(4,-2,-2)]_H = \begin{pmatrix} a \\ b \\ c \end{pmatrix}$$

where

$$(4,-2,-2) = a(1,0,0) + b(1,-1,0) + c(1,0,-1)$$

Similarly,

$$[(4,-3,-2)]_H = \begin{pmatrix} p \\ q \\ r \end{pmatrix}$$

where

$$(4,-3,-2) = P(1,0,0) + q(1,-1,0) + r(1,0,-1)$$

$$[(3,-2,-3)]_H = \begin{pmatrix} s \\ t \\ w \end{pmatrix}$$

where

$$(3,-2,-3) = s(1,0,0) + t(1,-1,0) + w(1,0,-1)$$

Each of these requires solving a system of linear equation in three unknows
Solving the first system, we have

$$4 = a + b + c$$

$$-2 = -b \;, -2 = -c$$

$$\Rightarrow a = 0 \;, b = c = 2$$

$$\Rightarrow [(4, -2, -2)]_H = \begin{pmatrix} 0 \\ 2 \\ 2 \end{pmatrix}$$

Solving the second system, we have

$$4 = p + q + r$$

$$-3 = -q \;, -2 = -r$$

$$\Rightarrow p = -1 \;, q = 3 \;, r = 2$$

$$\Rightarrow [(4, -3, -2)]_H = \begin{pmatrix} -1 \\ 3 \\ 2 \end{pmatrix}$$

Solving the third system, we have

$$3 = s + t + w$$

$$-2 = -t \;, -3 = -w$$

$$\Rightarrow s = -2 \;, t = 2 \;, w = 3$$

$$\Rightarrow [(3, -2, -3)]_H = \begin{pmatrix} -2 \\ 2 \\ 3 \end{pmatrix}$$

Therefore, the matrix representation of T with respect to basis G is

$$[T]_G^G = [T]_G = \begin{pmatrix} 0 & -1 & -2 \\ 2 & 3 & 2 \\ 2 & 2 & 3 \end{pmatrix}$$

Let λ be an eigenvalue of $[T]_G$, then

$$|\lambda I - [T]_G| = 0$$

$$\Rightarrow \begin{vmatrix} \lambda & 1 & 2 \\ -2 & \lambda - 3 & -2 \\ -2 & -2 & \lambda - 3 \end{vmatrix} = 0$$

$$(R_1 + R_2)$$

210

$$\Rightarrow \begin{vmatrix} \lambda-2 & \lambda-2 & 0 \\ -2 & \lambda-3 & -2 \\ -2 & -2 & \lambda-3 \end{vmatrix} = 0$$

Expanding along the first row, we have

$$(\lambda-2)((\lambda-3)^2 - 4) - (\lambda-2)(-2(\lambda-3) - 4) = 0$$
$$\Rightarrow (\lambda-2)(\lambda-3-2)(\lambda-3+2) + 2(\lambda-2)(\lambda-1) = 0$$
$$\Rightarrow (\lambda-1)(\lambda-2)(\lambda-5+2) = 0$$
$$\Rightarrow (\lambda-1)(\lambda-2)(\lambda-3) = 0$$

$\Rightarrow \lambda = 1,2,3$ – confirming the fact that similar matrices have the same eigenvalues.

b) for the eigenvectors of $[T]_E$, we have

$$(\lambda I - [T]_E)x = 0$$

For $\lambda = 1$, we have

$$\begin{pmatrix} -3 & 0 & -1 \\ 2 & 0 & 0 \\ 2 & 0 & 0 \end{pmatrix}\begin{pmatrix} x_1 \\ x_2 \\ x_3 \end{pmatrix} = \begin{pmatrix} 0 \\ 0 \\ 0 \end{pmatrix}$$

$$\Rightarrow 2x_1 = 0, x_2 = s, -3x_1 - x_3 = 0$$
$$\Rightarrow x_1 = 0, x_2 = s, x_3 = 0$$

$$\Rightarrow x = \begin{pmatrix} x_1 \\ x_2 \\ x_3 \end{pmatrix} = \begin{pmatrix} 0 \\ s \\ 0 \end{pmatrix} = s\begin{pmatrix} 0 \\ 1 \\ 0 \end{pmatrix}$$

$\Rightarrow$ The eigenvector of A corresponding to $\lambda = 1$ is $x = \begin{pmatrix} 0 \\ 1 \\ 0 \end{pmatrix}$.

The basis for the corresponding eigenspace of $[T]_E$ for $\lambda = 1$ is $G = \{(0,1,0)\}$

For $\lambda = 2$, we have

$$\begin{pmatrix} -2 & 0 & -1 \\ 2 & 1 & 0 \\ 2 & 0 & 1 \end{pmatrix}\begin{pmatrix} x_1 \\ x_2 \\ x_3 \end{pmatrix} = \begin{pmatrix} 0 \\ 0 \\ 0 \end{pmatrix}$$

$$\begin{pmatrix} -2 & 0 & -1 & | & 0 \\ 2 & 1 & 0 & | & 0 \\ 2 & 0 & 1 & | & 0 \end{pmatrix} (R_3 + R_1) \begin{pmatrix} -2 & 0 & -1 & | & 0 \\ 2 & 1 & 0 & | & 0 \\ 2 & 0 & 1 & | & 0 \end{pmatrix}$$

$$(R_2 + R_1) \begin{pmatrix} -2 & 0 & -1 & | & 0 \\ 0 & 1 & -1 & | & 0 \\ 0 & 0 & 0 & | & 0 \end{pmatrix}$$

$$\Rightarrow x_2 = x_3, \qquad -2x_1 = x_3$$

$$\Rightarrow x = \begin{pmatrix} x_1 \\ x_2 \\ x_3 \end{pmatrix} = \begin{pmatrix} \frac{-1}{2} x_3 \\ x_3 \\ x_3 \end{pmatrix} = \frac{1}{2} x_3 \begin{pmatrix} -1 \\ 2 \\ 2 \end{pmatrix} \text{for all real } x_3.$$

Therefore the eigenvalue of $[T]_E$ corresponding to $\lambda = 2$ is $x = \begin{pmatrix} -1 \\ 2 \\ 2 \end{pmatrix}$. For the

eigenspace of $[T]_E$ corresponding to $\lambda = 2$, the basis is $G = \{(-1, 2, 2)\}$.

For $\lambda = 3$, we have

$$\begin{pmatrix} -1 & 0 & -1 \\ 2 & 2 & 0 \\ 2 & 0 & 2 \end{pmatrix} \begin{pmatrix} x_1 \\ x_2 \\ x_3 \end{pmatrix} = \begin{pmatrix} 0 \\ 0 \\ 0 \end{pmatrix}$$

$$\Rightarrow x_1 = -x_3, x_1 = -x_2$$

$$\Rightarrow x = \begin{pmatrix} x_1 \\ x_2 \\ x_3 \end{pmatrix} = \begin{pmatrix} x_1 \\ -x_1 \\ -x_1 \end{pmatrix} = x_1 \begin{pmatrix} 1 \\ -1 \\ -1 \end{pmatrix} = -x_1 \begin{pmatrix} -1 \\ 1 \\ 1 \end{pmatrix} \text{ for all real } x_1.$$

Therefore, the eigenvector of $[T]_E$ corresponding to $\lambda = 3$ is

$$x = \begin{pmatrix} 1 \\ -1 \\ -1 \end{pmatrix} or \ x = \begin{pmatrix} -1 \\ 1 \\ 1 \end{pmatrix}, \text{ but not both.}$$

The eigenspace of $[T]_E$ corresponding to $\lambda = 3$ has the basis

$$G = \{(-1, 1, 1)\} \ or \ G = \{(1, -1, -1)$$

Now we consider $[T]_H$. For $\lambda = 1$, we have

$$\begin{pmatrix} 1 & 1 & 2 \\ -2 & -2 & -2 \\ -2 & -2 & -2 \end{pmatrix} \begin{pmatrix} x_1 \\ x_2 \\ x_3 \end{pmatrix} = \begin{pmatrix} 0 \\ 0 \\ 0 \end{pmatrix}$$

$$\begin{pmatrix} 1 & 1 & 2 & | & 0 \\ -2 & -2 & -2 & | & 0 \\ -2 & -2 & -2 & | & 0 \end{pmatrix} (R_3 - R_2), (R_2 + 2R_1) \begin{pmatrix} 1 & 1 & 2 & | & 0 \\ 0 & 0 & 2 & | & 0 \\ 0 & 0 & 0 & | & 0 \end{pmatrix}$$

$$(R_1 - R_2) \begin{pmatrix} 1 & 1 & 0 & | & 0 \\ 0 & 0 & 2 & | & 0 \\ 0 & 0 & 0 & | & 0 \end{pmatrix}$$

$$\Rightarrow \quad 2x_3 = 0, \; x_1 = -x_2$$

$$\Rightarrow \quad x = \begin{pmatrix} x_1 \\ x_2 \\ x_3 \end{pmatrix} = \begin{pmatrix} x_1 \\ -x_1 \\ 0 \end{pmatrix} = x_1 \begin{pmatrix} 1 \\ -1 \\ 0 \end{pmatrix}$$

$$or$$

$$x = \begin{pmatrix} -x_2 \\ x_2 \\ 0 \end{pmatrix} = x_2 \begin{pmatrix} -1 \\ 1 \\ 0 \end{pmatrix}$$

$\Rightarrow$ The eigenvector of $[T]_H$ corresponding to $\lambda = 1$ is $x = \begin{pmatrix} 1 \\ -1 \\ 0 \end{pmatrix}$. This means that

the corresponding eigenspace has a basis $S = \{(1, -1, 0)\}$.

For $\lambda = 2$, we have

$$\begin{pmatrix} 2 & 1 & 2 \\ -2 & -1 & -2 \\ -2 & -2 & -1 \end{pmatrix} \begin{pmatrix} x_1 \\ x_2 \\ x_3 \end{pmatrix} = \begin{pmatrix} 0 \\ 0 \\ 0 \end{pmatrix}$$

$$\begin{pmatrix} 2 & 1 & 2 & | & 0 \\ -2 & -1 & -2 & | & 0 \\ -2 & -2 & -1 & | & 0 \end{pmatrix} (R_2 + R_1)(R_3 + R_1) \begin{pmatrix} 2 & 1 & 2 & | & 0 \\ 0 & 0 & 0 & | & 0 \\ 0 & -1 & 1 & | & 0 \end{pmatrix}$$

$$(R_1 + R_3) \begin{pmatrix} 2 & 0 & 3 & | & 0 \\ 0 & 0 & 0 & | & 0 \\ 0 & -1 & 1 & | & 0 \end{pmatrix}$$

$$\Rightarrow x_2 = x_3, \; 2x_1 = -3x_3$$

$$\Rightarrow \quad x = \begin{pmatrix} x_1 \\ x_2 \\ x_3 \end{pmatrix} = \begin{pmatrix} -\frac{3}{2}x_3 \\ x_3 \\ x_3 \end{pmatrix} = \frac{-1}{2}x_3 \begin{pmatrix} 3 \\ -2 \\ -2 \end{pmatrix} \text{ for all real } x_3.$$

$\Rightarrow$ The eigenvector of $[T]_H$ corresponding to $\lambda = 2$ is $x = \begin{pmatrix} 3 \\ -2 \\ -2 \end{pmatrix}$. Therefore

the corresponding eigenspace has the basis $S = \{(3, -2, -2)\}$.

For $\lambda = 3$, we have

$$\begin{pmatrix} 3 & 1 & 2 \\ -2 & 0 & -2 \\ -2 & -2 & 0 \end{pmatrix}\begin{pmatrix} x_1 \\ x_2 \\ x_3 \end{pmatrix} = \begin{pmatrix} 0 \\ 0 \\ 0 \end{pmatrix}$$

$$\Rightarrow \quad -2x_1 = 2x_2, -2x_1 = 2x_3$$

$$\Rightarrow x = \begin{pmatrix} x_1 \\ x_2 \\ x_3 \end{pmatrix} = \begin{pmatrix} x_1 \\ -x_1 \\ -x_1 \end{pmatrix} = x_1\begin{pmatrix} 1 \\ -1 \\ -1 \end{pmatrix} \text{ or } s\begin{pmatrix} -1 \\ 1 \\ 1 \end{pmatrix}, \text{ where } s \text{ is a real number.}$$

Therefore, the eigenvector corresponding to $\lambda = 3$ for $[T]_H$ is $x = \begin{pmatrix} 1 \\ -1 \\ -1 \end{pmatrix}$, and

the corresponding eigenspace has the basis $S = \{(1, -1, -1)\}$ or

$S = \{(-1, 1, 1)\}$.

To confirm our results, recall that $[T]_E$ and $[T]_H$ are similar. This means that there is a nonsingular P such that

$$[T]_E = P[T]_H P^{-1}$$

In this case, P is the change of basis matrix from basis H to basis E. Indeed,

$P = \begin{pmatrix} 1 & 1 & 1 \\ 0 & -1 & 0 \\ 0 & 0 & -1 \end{pmatrix}$, the columns are the basis vectors of H. We have already

established that if

$$A = PA'P^{-1}$$

And x is an eigenvector of A, then $P^{-1}x$ is an eigenvector of A'. This implies that if x is an eigenvector for $[T]_E$, then $P^{-1}x$ is an eigenvector for $[T]_H$. Let us quickly find P^{-1}.

$$P = \begin{pmatrix} 1 & 1 & 1 & | & 1 & 0 & 0 \\ 0 & -1 & 0 & | & 0 & 1 & 0 \\ 0 & 0 & -1 & | & 0 & 0 & 1 \end{pmatrix} \quad (R_1 + R_2) \quad \begin{pmatrix} 1 & 0 & 1 & | & 1 & 1 & 0 \\ 0 & -1 & 0 & | & 0 & 1 & 0 \\ 0 & 0 & -1 & | & 0 & 0 & 1 \end{pmatrix}$$

$$(R_1 + R_3) \begin{pmatrix} 1 & 0 & 0 & | & 1 & 1 & 1 \\ 0 & -1 & 0 & | & 0 & 1 & 0 \\ 0 & 0 & -1 & | & 0 & 0 & 1 \end{pmatrix} (-R_2), (-R_2) \begin{pmatrix} 1 & 0 & 0 & | & 1 & 1 & 1 \\ 0 & 1 & 0 & | & 0 & -1 & 0 \\ 0 & 0 & 1 & | & 0 & 0 & -1 \end{pmatrix}$$

$$P^{-1} = \begin{pmatrix} 1 & 1 & 1 \\ 0 & -1 & 0 \\ 0 & 0 & -1 \end{pmatrix}$$

Now let's test.

For $x_E = \begin{pmatrix} 0 \\ 1 \\ 0 \end{pmatrix}$

$$x_H = P^{-1} x_E$$

$$\Rightarrow \qquad x_H = \begin{pmatrix} 1 & 1 & 1 \\ 0 & -1 & 0 \\ 0 & 0 & -1 \end{pmatrix} \begin{pmatrix} 0 \\ 1 \\ 0 \end{pmatrix} = \begin{pmatrix} 1 \\ -1 \\ 0 \end{pmatrix}, \text{ which was what we got.}$$

For $x_E = \begin{pmatrix} -1 \\ 2 \\ 2 \end{pmatrix}$,

$$x_H = P^{-1} x_E = \begin{pmatrix} 1 & 1 & 1 \\ 0 & -1 & 0 \\ 0 & 0 & -1 \end{pmatrix} \begin{pmatrix} -1 \\ 2 \\ 2 \end{pmatrix} = \begin{pmatrix} 3 \\ -2 \\ -2 \end{pmatrix}, \text{ also confirming our}$$

previous result.

Finally, for

$$x_E = \begin{pmatrix} 1 \\ -1 \\ -1 \end{pmatrix},$$

$$x_H = P^{-1} x_E = \begin{pmatrix} 1 & 1 & 1 \\ 0 & -1 & 0 \\ 0 & 0 & -1 \end{pmatrix} \begin{pmatrix} 1 \\ -1 \\ -1 \end{pmatrix} = \begin{pmatrix} -1 \\ 1 \\ 1 \end{pmatrix}$$

or

For $x_E = \begin{pmatrix} -1 \\ 1 \\ 1 \end{pmatrix}$,

$$x_H = P^{-1}x_E = \begin{pmatrix} 1 & 1 & 1 \\ 0 & -1 & 0 \\ 0 & 0 & -1 \end{pmatrix}\begin{pmatrix} -1 \\ 1 \\ 1 \end{pmatrix} = \begin{pmatrix} 1 \\ -1 \\ -1 \end{pmatrix}$$

SOME RESULTS ON EIGENVALUES

Let A be an nxn matrix whose eigenvalues are $\lambda_1, \lambda_2, \ldots \ldots \lambda_n$, all not necessarily distinct, then

$$\prod_{i=1}^{n} \lambda_i = \lambda_1 \lambda_2 \ldots \lambda_n = |A|$$

Proof

Recall that if λ is an eigenvalue of A, then

$$|\lambda I - A| = 0$$

From our previous work, we also know that $|\lambda I - A|$ is always a polynomial of degree n in λ which can also be uniquely factorized (although not always easy) as

$$(\lambda - \lambda_1)(\lambda - \lambda_2) \ldots \ldots (\lambda - \lambda_n)$$

Where $\lambda_1, \lambda_2, \ldots \ldots \lambda_n$ are eigenvalues of A.

$$\Rightarrow (\lambda - \lambda_1)(\lambda - \lambda_2) \ldots \ldots (\lambda - \lambda_n) = |\lambda I - A|$$

In particular, any real number substituted for λ on both sides of this equation does not alter the equation. Therefore, putting zero for λ, we have

$$(-\lambda_1)(\lambda_2) \ldots \ldots (-\lambda_n) = |-A|$$

$\Rightarrow \quad (-1)^n \lambda_1 \lambda_2 \ldots \ldots \lambda_n = (-1)^n |A|$; recall that $|kA| = k^n|A|$ for any real k.

$$\Rightarrow \quad \lambda_1 \lambda_2 \ldots \ldots \lambda_n = |A| \qquad\qquad ----- P206$$

You can confirm this result from all our previous examples.

Let A be an nxn matrix whose eigenvalues are $\lambda_1, \lambda_2, \ldots \ldots \lambda_n$, all not necessarily distinct, then

$$\sum_{i=1}^{n} \lambda_i = trace(A). \qquad\qquad -----P207$$

Proof

It is a well known fact that

$$Trace\ (AB) = Trace(BA)$$

Again, be informed that any diagonalizable matrix $A(nxn)$ is always expressible as

$A = PDP^{-1}$, where D is a diagonal matrix whose diagonal elements are the eigenvalues of A, and P is an nxn invertible matrix whose *ith* column is the eigenvector of A corresponding to the eigenvalue d_{ii} element of D.

$$\Rightarrow \qquad Tr(A) = Tr(PDP^{-1})$$
$$= Tr(P(DP^{-1})) = Tr((DP^{-1})P)$$
$$= Tr(DPP^{-1})$$
$$= Tr(DI) = Tr(D)$$
$$\Rightarrow Tr(A) = \sum_{i=1}^{n} d_{ii} = \sum_{i=1}^{n} \lambda_i$$

Let A be an nxn matrix, then A and A^T have the same set of eigenvalues. In other words, λ is an eigenvalue of A^T whenever λ is an eigenvalue of A.

Proof

If λ is an eigenvalue of A corresponding to the eigenvector $\boldsymbol{x}$, then

$$A\boldsymbol{x} = \lambda\boldsymbol{x}$$
$$\Rightarrow (A\boldsymbol{x})^T \boldsymbol{x} = (\lambda\boldsymbol{x})^T \boldsymbol{x}$$
$$\Rightarrow (A\boldsymbol{x})^T \boldsymbol{x} = \lambda\,\boldsymbol{x}^T \boldsymbol{x}$$

But

$$(Ax)^T = x^T A^T$$

$$\Rightarrow \quad x^T A^T x = x^T \lambda x$$

$$x^T (A^T x - \lambda x) = 0$$

Since $x \neq 0$, $x^T \neq 0$

$$\Rightarrow \quad A^T x - \lambda x = 0$$

$$\therefore \quad A^T x = \lambda x \qquad\qquad -----P208$$

This implies that A and A^T have equal eigenvalues.

OR

Proof

Using the fact that

$$(\lambda I - A)^T = \lambda I - A^T = since \ (A + B)^T = A^T + B^T, \qquad I^T = I$$

$$\Rightarrow \quad |(\lambda I - A)^T| = |\lambda I - A^T|$$

But

$$|(\lambda I - A)^T| = |\lambda I - A| \ ; that \ is \ |A^T| = |A|$$

$$\Rightarrow |\lambda I - A| = |\lambda I - A^T| \qquad ----- P209$$

Let A be an nxn matrix, if λ is an eigenvalue of A, then $\frac{1}{\lambda}$ is an eigenvalue of A^{-1} provided that $|A| \neq 0$.

Proof

Let x be the corresponding eigenvector for λ, then

$$Ax = \lambda x$$

Premultiplying both sides with A^{-1}, we have

$$A^{-1}Ax = A^{-1}\lambda x$$

$$\Rightarrow \quad Ix = A^{-1}\lambda x$$

$$\Rightarrow \frac{1}{\lambda} I\boldsymbol{x} = A^{-1}\boldsymbol{x}$$

$$\Rightarrow A^{-1}\boldsymbol{x} = \frac{1}{\lambda}\boldsymbol{x} \qquad - - - - - P210$$

Thus $\frac{1}{\lambda}$ is an eigenvalue of A^{-1} corresponding to the eigenvector $\boldsymbol{x}$.

Let A be an nxn matrix, if λ is an eigenvalue of A, then for any nonzero scalar $k, k\lambda$ is an eigenvalue of kA.

Proof

Let $\boldsymbol{x}$ be the corresponding eigenvector for the eigenvalue λ, then

$$A\boldsymbol{x} = \lambda\boldsymbol{x}$$

Multiplying both sides by k, we have

$$kA\boldsymbol{x} = k\lambda\boldsymbol{x}$$

$$\Rightarrow (kA)\boldsymbol{x} = (k\lambda)\boldsymbol{x} \qquad - - - - - P211$$

$\Rightarrow k\lambda$ is an eigenvalue of kA whenever λ is an eigenvalue of A; also, they have the same eigenvectors.

Let A be an nxn matrix, if λ is an eigenvalue of A, then for any positive interger m, λ^m is an eigenvalue of A^m.

Proof

We prove by induction

$A\boldsymbol{x} = \lambda\boldsymbol{x}$; $\boldsymbol{x}$ is the corresponding eigenvector. This implies that it holds for $m = 1$.

Suppose that it holds for $m = k$, a positive integer, then

$$A^k\boldsymbol{x} = \lambda^k\boldsymbol{x}$$

$$\Rightarrow AA^k\boldsymbol{x} = A\lambda^k\boldsymbol{x}$$

$$\Rightarrow A^{k+1}\boldsymbol{x} = \lambda^k A\boldsymbol{x}$$

$$\Rightarrow A^{k+1}\boldsymbol{x} = \lambda^k \lambda\boldsymbol{x} ; \; since \; A\boldsymbol{x} = \lambda\boldsymbol{x}$$

$$\Rightarrow A^{k+1}\boldsymbol{x} = \lambda^{k+1}\boldsymbol{x}$$

$$\Rightarrow A^m x = \lambda^m x \qquad\qquad ----- P212$$

Therefore, it holds for all positive integers. Moreover, A and A^m have the same eigenvectors.

Let both A and B be nxn matrix, then AB and BA have equal eigenvalues.

Proof

Let λ be an eigenvalue of AB corresponding to the eigenvector x then

$$(AB)x = \lambda x$$

It is not difficult to see that BA and AB are similar because

$$AB = (B^{-1}B)AB$$
$$\Rightarrow AB = B^{-1}(BA)B$$

which shows that AB and BA are similar. We have already shown that similar matrices have equal eigenvalues.

$$\therefore (AB)x = \lambda x \Rightarrow (BA)x = \lambda x \qquad\qquad ------ P213$$

Let A be an nxn matrix whose eigenvalue is λ, then for any polynomial

$$f(x) = \sum_{k=0}^{m} a_k x^k \text{ , the eigenvalue of } (If(A)) \text{ is } f(\lambda).$$

Proof

We have already established that if λ is an eigenvalue of A, then

$$Ax = \lambda x$$
$$\Rightarrow a_k Ax = a_k \lambda x \text{ for real } a_k.$$

Recall that $A^k x = \lambda^k x$ for any nonnegative k because $Ix = \lambda^0 x$

$$\Rightarrow a_k A^k x = a_k \lambda^k x$$
$$\Rightarrow \sum_{k=0}^{m} a_k A^k x = \sum_{k=0}^{m} a_k \lambda^k x$$

$$\Rightarrow \left(I \sum_{k=0}^{m} a_k A^k \right) x = \left(\sum_{k=0}^{m} a_k \lambda^k \right) x$$

$$\Rightarrow \quad If(A)x = f(\lambda)x \qquad -----P214$$

If A, an nxn matrix, is a diagonal matrix, then the eigenvalues of A are the diagonal elements of A.

Proof

$$A = \begin{pmatrix} d_1 & 0 & 0\ldots\ldots & \ldots \\ 0 & d_2 & \ldots\ldots\ldots & \vdots \\ 0 & \vdots & \ddots & \vdots \\ 0 & \vdots & \ddots & \vdots \\ \vdots & \vdots & \ldots\ldots\ldots & d_n \end{pmatrix}$$

$$|\lambda I - A| = 0$$

$$\Rightarrow |\lambda I - A| = \begin{vmatrix} \lambda - d_1 & 0 & 0\ldots\ldots & 0 \\ 0 & \lambda - d_2 & \ldots\ldots\ldots & \vdots \\ 0 & \vdots & \ddots & \vdots \\ \vdots & \vdots & \ddots & \vdots \\ 0 & 0 & \ldots\ldots\ldots & \lambda - d_n \end{vmatrix} = 0$$

$$\Rightarrow \quad (\lambda - d_1)(\lambda - d_2)\ldots\ldots(\lambda - d_n) = 0$$

$$\Rightarrow \quad \lambda = d_1, d_2, \ldots d_n \qquad -----P215$$

221

Cayley- Hamilton's Theorem

Every square matrix satisfies its own characteristics polynomial.

Proof

It is already established that given any nxn matrix A whose eigenvalue is λ, then for any polynomial

$$f(x) = \sum_{k=0}^{m} a_k x^k \, ,$$

$I(f(A)x) = f(\lambda)x, \ where \ x$ is the corresponding eigenvector for λ.

Now, in particular, if we select $f(\lambda)$ to be the characteristics polynomial of A, then

$$f(\lambda) = 0$$

$$\Rightarrow I(f(A)x) = \mathbf{0}, the \ null \ nx1 \ vector.$$

Since $x \neq 0$

$$\Rightarrow I\big(f(A)\big) = 0$$

where $f(\lambda)$ is the characteristics polynomial of A. Multiplying by I is necessary because the constant term of the polynomial has to be in matrix form for the equation to be meaningful, since it is meaningless to add matrix to a scalar. In the last equation, 0 represents the null nxn matrix.

Example5.6

For $0< t < \pi$, the matrix $A = \begin{pmatrix} \cos t & -\sin t \\ \sin t & \cos t \end{pmatrix}$ has distinct complex eigenvalues λ_1 and λ_2. For what value of t, $0< t < \pi$, is $\lambda_1 + \lambda_2 = 1$?

Solution

It has been established already that sum of eigenvalues is equal to the trace, and the product of eigenvalues is equal to the determinant.

$$\therefore \ \lambda_1 + \lambda_2 = trace(A)$$

$$\Rightarrow \ 2Cost = 1$$

$$t = \cos^{-1}\left(\frac{1}{2}\right) = \frac{\pi}{3}$$

Example 5.7

Let A be a 2x2 matrix for which there is a constant k such that the sum of the entries in each row and in each column is k. Find the eigenvector of A.

Solution

From the description of A given,

$$A = \begin{pmatrix} k-c & c \\ c & k-c \end{pmatrix}$$

If λ is an eigenvalue of A, then

$$\begin{vmatrix} k-c & c \\ c & k-c \end{vmatrix} = 0$$

$$\Rightarrow (\lambda - c)^2 - (c-k)^2 = 0$$

$$\Rightarrow \lambda - c = \pm(c-k)$$

$$\Rightarrow \lambda = 2c - k \text{ or } \lambda = k$$

When $\lambda = k$, we have

$$\begin{pmatrix} k-c & c-k \\ c-k & k-c \end{pmatrix}\begin{pmatrix} x_1 \\ x_2 \end{pmatrix} = \begin{pmatrix} 0 \\ 0 \end{pmatrix}$$

$$\Rightarrow (k-c)x_1 = -(c-k)x_2$$

$$\Rightarrow x_1 = x_2 = s \text{ ; } s \text{ is any real number.}$$

$$\Rightarrow x = \begin{pmatrix} s \\ s \end{pmatrix} = s\begin{pmatrix} 1 \\ 1 \end{pmatrix}$$

Therefore $x = \begin{pmatrix} 1 \\ 1 \end{pmatrix}$ is an eigenvector of A.

When $\lambda = 2c - k$, we have

$$\begin{pmatrix} c-k & c-k \\ c-k & c-k \end{pmatrix}\begin{pmatrix} x_1 \\ x_2 \end{pmatrix} = \begin{pmatrix} 0 \\ 0 \end{pmatrix}$$

$$\Rightarrow (c-k)x_1 = -(c-k)x_2$$

$$\Rightarrow x_1 = -x_2$$

$$\Rightarrow x = \begin{pmatrix} x_1 \\ x_2 \end{pmatrix} = \begin{pmatrix} x_1 \\ -x_1 \end{pmatrix} = x_1\begin{pmatrix} 1 \\ -1 \end{pmatrix}$$

$$or$$

$$\Rightarrow \; x = \begin{pmatrix} -x_2 \\ x_2 \end{pmatrix} = x_2 \begin{pmatrix} -1 \\ 1 \end{pmatrix}$$

Therefore, the eigenvectors of A are

$$x = \begin{pmatrix} 1 \\ 1 \end{pmatrix} \text{ and } \begin{pmatrix} 1 \\ -1 \end{pmatrix} \text{ or } x = \begin{pmatrix} 1 \\ 1 \end{pmatrix} \text{ and } \begin{pmatrix} -1 \\ 1 \end{pmatrix}$$

- *Let A be an nxn matrix with distinct eigenvalues λ_1 and λ_2. If x is an eigenvector of A corresponding to λ_1, then x cannot correspond to λ_2.*

Proof

x is an eigenvector corresponding to the eigenvalue λ_1 of A

$$\Rightarrow \; Ax = \lambda_1 x$$

Suppose that x also corresponds to λ_2, then

$$Ax = \lambda_2 x$$

$$\Rightarrow Ax - Ax = \lambda_2 x - \lambda_1 x$$

$$\Rightarrow (\lambda_2 - \lambda_1)x = 0$$

Since $x \neq 0$

$$\Rightarrow \lambda_2 - \lambda_1 = 0$$

$\Rightarrow \lambda_2 = \lambda_1$, a contradiction to our initial assumption that $\lambda_1 \neq \lambda_2$. Therefore x cannot correspond to λ_2 if it does to λ_1.

SIMILARITY TO DIAGONAL MATRICES: DIAGONALIZATION

We have already shown that a linear transformation T from the finite – dimensional vector space V into itself can be represented, relative to a given basis for V, as a matrix multiplication. The precise matrix representation of T is dependent on the choice of the basis for V; matrix representation for T varies with the basis for V. We have also seen that all such matrix representations for T are similar, and thus have equal eigenvalues.

Now we shift our focus to determining under what conditions a basis for V can be chosen so that the matrix A representing T is a diagonal matrix.

A square matrix A is said to be **diagonalizable** if there exists an invertible matrix P such that the matrix $D = P^{-1}AP$ is a diagonal matrix. In other words a square matrix A is diagonalizable if and only if A is similar to a diagonal matrix. In fact, the word "*diagonalization*" means "**similarity to a diagonal matrix**"

* *An nxn matrix A is similar to a diagonal matrix D if and only if A has n linearly independent eigenvectors. Moreover, the diagonal elements of D are the common eigenvalues of A and D.*

Proof

Assume that $P^{-1}AP = D$, a diagonal matrix, then

$$AP = PD. \text{ Let}$$

$$P = \begin{pmatrix} p_{11} & p_{12} & \cdots\cdots\cdots & p_{1n} \\ p_{12} & p_{22} & \cdots\cdots\cdots & p_{2n} \\ \vdots & \vdots & & \vdots \\ p_{n1} & p_{n2} & \cdots\cdots\cdots & p_{nn} \end{pmatrix} \text{ and } D = \begin{pmatrix} d_{11} & 0 & \cdots\cdots\cdots & 0 \\ 0 & d_{22} & \cdots\cdots\cdots & 0 \\ \vdots & \vdots & \ddots & \vdots \\ \vdots & \vdots & & \ddots & \vdots \\ 0 & 0 & \cdots\cdots\cdots & d_{nn} \end{pmatrix}$$

$$\text{Then } AP = PD = \begin{pmatrix} d_{11}p_{11} & d_{22}p_{12} & \cdots\cdots\cdots & d_{nn}p_{1n} \\ d_{11}p_{21} & d_{22}p_{22} & \cdots\cdots\cdots & d_{nn}p_{2n} \\ \vdots & \vdots & & \vdots \\ d_{11}p_{n1} & d_{22}p_{n2} & \cdots\cdots\cdots & d_{nn}p_{nn} \end{pmatrix}$$

Let the ith column vector of P be denoted as $\boldsymbol{p}_i$, then the above matrix equation can be written as

$$A\boldsymbol{p}_i = d_{ii}\boldsymbol{p}_i, i = 1, 2, \ldots\ldots n$$

Since none of the $\boldsymbol{p}_i$ is the zero vector because P is invertible (by definition), then

$$A\boldsymbol{p}_i = d_{ii}\boldsymbol{p}_i$$

$\Rightarrow \boldsymbol{p}_i$ is an eigenvector of A corresponding to the eigenvalue d_{ii} for each i. Moreover, the column vectors of P are linearly independent since $|P| \neq 0$.

Conversely, if the matrix A has n linearly independent eigenvectors $\boldsymbol{p}_1, \boldsymbol{p}, \dots \dots \boldsymbol{p}_n$ corresponding to the eigenvalues $\lambda_1, \lambda_2, \dots \dots \lambda_n$, then the matrix P constructed as

$P = (\boldsymbol{p}_1 : \boldsymbol{p}_2 : \dots \dots \dots \boldsymbol{p}_n)$ is invertible because its column vectors $\boldsymbol{p}_1, \boldsymbol{p}_2, \dots \dots \boldsymbol{p}_n$ are linearly independent. Thus,

$$AP = (A\boldsymbol{p}_1 : A\boldsymbol{p}_2 : \dots \dots \dots A\boldsymbol{p}_n)$$

$$= (\lambda_1 \boldsymbol{p}_1 : \lambda_2 \boldsymbol{p}_2 : \dots \dots \dots \lambda_n \boldsymbol{p}_n)$$

$$= P \begin{pmatrix} \lambda_1 & 0 & \dots \dots \dots & 0 \\ 0 & \lambda_2 & \dots \dots \dots & 0 \\ \vdots & \vdots & \ddots & \vdots \\ \vdots & \vdots & \ddots & \vdots \\ 0 & 0 & \dots \dots \dots & \lambda_n \end{pmatrix}$$

$$\Rightarrow AP = PD$$

Hence,

$$D = P^{-1}AP$$

and A is similar to a diagonal matrix with its eigenvalues on the main diagonal.

Let us see how to construct the matrix P by considering the following examples

Example5.8

Diagonalize, if possible, the matrix

$A = \begin{pmatrix} 3 & 1 \\ 1 & 3 \end{pmatrix}$

Solution

We have already seen that the eigenvalues of A are $\lambda = 4$ and $\lambda = 2$, and there corresponding eigenvectors are $x = \begin{pmatrix} 1 \\ 1 \end{pmatrix}$ and $x = \begin{pmatrix} 1 \\ -1 \end{pmatrix}$. Therefore, the matrix P that diagonalizes A is

$$P = \begin{pmatrix} 1 & 1 \\ 1 & -1 \end{pmatrix}, \text{ the eigenvectors form the columns.}$$

Also,

$$D = \begin{pmatrix} 4 & 0 \\ 0 & 2 \end{pmatrix}, \text{ the eigenvalues form the diagonal elements.}$$

$$\Rightarrow D = P^{-1}AP$$

$$\Rightarrow \begin{pmatrix} 4 & 0 \\ 0 & 2 \end{pmatrix} = \begin{pmatrix} 1 & 1 \\ 1 & -1 \end{pmatrix}^{-1} \begin{pmatrix} 3 & 1 \\ 1 & 3 \end{pmatrix} \begin{pmatrix} 1 & 1 \\ 1 & -1 \end{pmatrix}$$

You can check this by compressing the RHS to a single matrix.

Example 5.9

Diagonalize, if possible, the matrix

$$A = \begin{pmatrix} 1 & -2 \\ 2 & -3 \end{pmatrix}$$

Solution

Let λ be an eigenvalue of A, then

$$|\lambda I - A| = 0$$

$$\Rightarrow \begin{vmatrix} \lambda - 1 & 2 \\ -2 & \lambda + 3 \end{vmatrix} = 0$$

$$\Rightarrow (\lambda - 1)(\lambda + 3) + 4 = 0$$

$$\Rightarrow \lambda^2 + 2\lambda + 1 = 0$$

$$\Rightarrow \lambda = -1, \text{ repeated}$$

For the eigenvalue, we have

$$\begin{pmatrix} -2 & 2 \\ -2 & 2 \end{pmatrix} \begin{pmatrix} x_1 \\ x_2 \end{pmatrix} = \begin{pmatrix} 0 \\ 0 \end{pmatrix}$$

$$x_1 = x_2 = s, \text{ where s is a scaler.}$$

$$s = \begin{pmatrix} x_1 \\ x_2 \end{pmatrix} = \begin{pmatrix} s \\ s \end{pmatrix} = s\begin{pmatrix} 1 \\ 1 \end{pmatrix}$$

$$\Rightarrow \text{ The only eigenvector is } x = \begin{pmatrix} 1 \\ 1 \end{pmatrix}.$$

This means that matrix A is not diagonalizable because it fails to have two linearly independent eigenvectors as required that an $n \times n$ matrix must have n linearly independent eigenvectors to be diagonalizable-similar to a diagonal matrix.

Note that matrix A is not similar to a diagonal matrix, not because it has only one eigenvalue but because the dimension of the eigenspace corresponding to this eigenvalue is one. We have seen in a previous example that an $n \times n$ matrix A can have k distinct eigenvalues $(k < n)$ and still have a linearly independent eigenvectors. Therefore, what is important is that, weather there are n distinct eigenvalues or not, the respective dimensions of all the eigenspaces must sum up to n. We illustrate this with the following example.

Example 5.10

Diagonalize, if possible, the matrix

$$A = \begin{pmatrix} 4 & 2 & 0 \\ 2 & 4 & 0 \\ 0 & 0 & 2 \end{pmatrix}$$

Solution

We have established already, in a previous example, that the eigenvalues of A are $\lambda = 2$, repeated and $\lambda = 6$

The corresponding eigenvalues for $\lambda = 2$ are

$$x = \begin{pmatrix} 1 \\ -1 \\ 0 \end{pmatrix} \text{ and } x = \begin{pmatrix} 0 \\ 0 \\ 1 \end{pmatrix}, \text{ and that for } \lambda = 6 \text{ is } x = \begin{pmatrix} 1 \\ 1 \\ 0 \end{pmatrix}.$$

We also established the linear independence of $\begin{pmatrix} 1 \\ -1 \\ 0 \end{pmatrix}$ and $\begin{pmatrix} 0 \\ 0 \\ 1 \end{pmatrix}$ although we did not show that the set $G = \{(1,-1,0), (0,0,1), (1,1,0)\}$ is linearly independent. It is very obvious that G is linearly independent. Therefore A is diagonalizable. The matrix P that diagonalizes A is

$$P = \begin{pmatrix} 1 & 0 & 1 \\ -1 & 0 & 1 \\ 0 & 1 & 0 \end{pmatrix}$$

$$\Rightarrow \qquad D = \begin{pmatrix} 2 & 0 & 0 \\ 0 & 2 & 0 \\ 0 & 0 & 6 \end{pmatrix} = P^{-1}AP$$

Be informed that eigenvectors corresponding to distinct eigenvalues are always linearly independent as will be shown later.

This fact will prevent us from attempting to establish the linear independence of such eigenvectors as seen in the last example. Again, note that a symmetric matrix is always diagonalizable; it is always similar to a diagonal matrix – meaning that even if a symmetric matrix $A(nxn)$ has $k < n$ distinct eigenvalues, the eigenvectors corresponding to the same eigenvalue are always linearly independent. For nonsymmetric matrices, the same thing may be observed, but it is not always so. This means that, in the case of non symmetric matrices, one needs to manually establish the linear independence of such eigenvectors belonging to the same eigenvalue.

Moreover, it should be noted that any matrix $A(nxn)$ which has n distinct eigenvalues is always similar to a diagonal matrix (diagonalizable) although this condition is not necessary; it is too demanding on A.

If we view a matrix as a representation of some linear transformation, indeed it is, on a finite – dimensional vector space V, then diagonalizing a matrix A means finding a particular basis of V that makes the matrix representation of the linear transformation a diagonal matrix. This basis is, indeed, the set of the linearly independent eigenvectors of the matrix. So, finding a matrix P that diagonalizes A means finding a basis for V such that A is a diagonal matrix; the diagonal elements are the eigenvalues of A. We illustrate this with the following example.

Example 5.11

Let $T: R^3 \to R^3$ be the transformation on the vector space P_2 of quadratic polynomials defined by

$$T(c + bx + ax^2) = 5c - (2a - 3b)x + (3a - 2b)x^2$$

a) Write a matrix representation for T relative to the standard basis

$$E = \{1, x, x^2\}$$

b) Find a basis for P_2 such that the matrix representation for T relative to that basis is a diagonal matrix.

Solution

a)
$$E = \{1, x, x^2\} \equiv \{(1,0,0), (0,1,0), (0,0,1)\}$$
$$\therefore T(1,0,0) = (5,0,0)$$
$$T(0,1,0) = (0,3,-2)$$
$$T(0,0,1) = (0,-2,3)$$
$$\Rightarrow [T]_E = \begin{pmatrix} 5 & 0 & 0 \\ 0 & 3 & -2 \\ 0 & -2 & 3 \end{pmatrix}$$

b) To find a new basis H for V such that $[T]_H$ is a diagonal matrix, we simply find the linearly indedpendent eigenvectors of $[T]_E$. Let λ be an eigenvalue of $[T]_E$, then

$$|\lambda I - [T]_E| = 0$$

$$\Rightarrow \begin{vmatrix} \lambda - 5 & 0 & 0 \\ 0 & \lambda - 3 & 2 \\ 0 & 2 & \lambda - 3 \end{vmatrix} = 0$$

Expanding along the first row or first column, we have

$$(\lambda - 5)((\lambda - 3)^2 - 4) = 0$$

$$\Rightarrow (\lambda - 5)(\lambda - 5)(\lambda - 1) = 0$$

$$\Rightarrow \lambda = 5, \text{ repeated, } \lambda = 1$$

We expect the eigenspace of $[T]_E$ corresponding to $\lambda = 5$ to have two linearly independent eigenvectors – to be of dimension two since symmetric matrices are always similar to a diagonal matrix. Therefore, we have for $\lambda = 5$

$$\begin{pmatrix} 0 & 0 & 0 \\ 0 & 2 & 2 \\ 0 & 2 & 2 \end{pmatrix} \begin{pmatrix} x_1 \\ x_2 \\ x_3 \end{pmatrix} = \begin{pmatrix} 0 \\ 0 \\ 0 \end{pmatrix}$$

$$\Rightarrow x_1 = s, x_2 = -x_3 = t \; ; \; s, t \text{ are any real numbers}$$

$$\Rightarrow x = \begin{pmatrix} x_1 \\ x_2 \\ x_3 \end{pmatrix} = \begin{pmatrix} s \\ t \\ -t \end{pmatrix} = s \begin{pmatrix} 1 \\ 0 \\ 0 \end{pmatrix} + t \begin{pmatrix} 0 \\ 1 \\ -1 \end{pmatrix}$$

It is obvious that the vectors $(1,0,0)$ and $(0,1,-1)$ are linearly independent. Therefore, the eigenspace of $[T]_E$ corresponding to the eigenvalue $\lambda = 5$ is of dimension two as expected.

For $\lambda = 1$, we have

$$\begin{pmatrix} -4 & 0 & 0 \\ 0 & -2 & 2 \\ 0 & 2 & -2 \end{pmatrix} \begin{pmatrix} x_1 \\ x_2 \\ x_3 \end{pmatrix} = \begin{pmatrix} 0 \\ 0 \\ 0 \end{pmatrix}$$

$$\Rightarrow x_1 = 0, x_2 = x_3 = t \; ; t \text{ is any scalar.}$$

$$\Rightarrow x = \begin{pmatrix} x_1 \\ x_2 \\ x_3 \end{pmatrix} = \begin{pmatrix} 0 \\ t \\ t \end{pmatrix} = t \begin{pmatrix} 0 \\ 1 \\ 1 \end{pmatrix}$$

Therefore the eigenvector corresponding to the eigenvalue $\lambda = 1$ is

$$x = \begin{pmatrix} 0 \\ 0 \\ 1 \end{pmatrix}$$

A basis for P_2 such that the matrix representation for T is diagonal is

$$H = \{(1,0,0), (0,1,-1), (0,1,1)\}$$

Moreover, this matrix representation of T relative to H is

$$[T]_H = \begin{pmatrix} 5 & 0 & 0 \\ 0 & 5 & 0 \\ 0 & 0 & 1 \end{pmatrix}$$

Later, we shall see some of the advantages of diagonalizing a diagonalizable matrix. For now, let's consider one of the advantages.

Power of a Matrix

- *Let A(nxn) be similar to a diagonal matrix D such that*

$$D = P^{-1}AP \text{ for some } |P| \neq 0, \text{ then}$$

$$A^k = PD^kP^{-1} \text{ for any nonnegative integer k.}$$

Proof

$$\text{If} \quad D = P^{-1}AP$$

$$\Rightarrow PDP^{-1} = A$$

We prove by induction. The fact that

$$A = PDP^{-1}$$

shows that it is already true for $k = 1$. It is true for $k = 0$ because $A^0 = D^0 = I$.

Now suppose that it is true for some positive integer m; that is

$$A^m = PD^mP^{-1}$$

$$\Rightarrow A.A^m = (PDP^{-1})PD^mP^{-1}$$

$$\Rightarrow A^{m+1} = PDP^{-1}PD^mP^{-1}$$

$$= PDD^mP^{-1}; \text{ since } P^{-1}P = I$$

$$= PD^{m+1}P^{-1}$$

Since the relation is true for $k = 1$, and it is also true for $k = m + 1$ whenever it is true for $k = m$, then it is true for all positive integers.

$$A^k = PD^k P^{-1} \qquad\qquad -----P216$$

The above formula makes it very easy to raise an nxn matrix A to any positive integer power. This is because if

$$D = diag\ (d_{11},\ d_{22} \ldots \ldots \ldots d_{nn}), \text{ then}$$

$D^k = diag\ (d_{11}^k,\ d_{22}^k, \ldots \ldots \ldots \ldots d_{nn,}^k)$. Please convince yourself of this; it is very obvious.

The above formula also proves very useful in obtaining the inverse of an nxn matrix A because it holds good for $k = -1$ because

$$A = PDP^{-1}$$
$$\Rightarrow A^{-1} = (PDP^{-1})^{-1}$$
$$\Rightarrow A^{-1} = ((PD)P^{-1})^{-1}$$
$$\Rightarrow A^{-1} = (P^{-1})^{-1}(PD)^{-1} \text{ , a property of inverse operation}$$
$$\Rightarrow A^{-1} = PD^{-1}P^{-1}$$

Again,

$$D = diag\ (d_{11,}d_{22,} \ldots \ldots \ldots \ldots \ldots d_{nn})$$

$\Rightarrow D^{-1} = diag\ \left(\frac{1}{d_{11,}},\frac{1}{d_{22,}} \ldots \ldots \frac{1}{d_{nn,}}\right)$, making it very easy to calculate A^{-1} provided that the eigenvalues are known. Please convince yourself that $D^{-1} = diag\ \left(\frac{1}{d_{11,}},\frac{1}{d_{22,}} \ldots \ldots \frac{1}{d_{nn}}\right)$. It is very easy; a single step of row-transformation method will quickly reveal it.

Before we take specific examples, let us consider another ingenious way of obtaining the inverse, and the positive powers of an nxn matrix A. Recall that every square matrix satisfies its characteristic polynomial according to the Caley Hamilton's theorem. That is, let the polynomial

$a_n\lambda^n + a_{n-1}\lambda^{n-1} + \cdots\cdots a_0 = |\lambda I - A|$ be the characteristic polynomial, then

$$a_n A^n + a_{n-1}A^{n-1} + \cdots\cdots\cdots a_o I = 0 \qquad ----- P217$$

Multiplying both sides by A^{-1}, we have

$$a_n A^{-1}A^n + a_{n-1}A^{-1}A^{n-1} + \cdots\cdots a_1 A^{-1}A + a_0 A^{-1} = 0$$

$$\Rightarrow a_0 A^{-1} = -a_n A^{n-1} - a_{n-1}A^{n-2} - \cdots\cdots a_1 I$$

$$\Rightarrow A^{-1} = \frac{-1}{a_o}(a_n A^{n-1} + a_{n-1}A^{n-2} + \cdots\cdots a_1 I) \qquad ----- P218$$

Multiplying P by A, we have

$$a_n A^{n+1} + a_{n-1}A^n + \ldots\ldots\ldots\ldots\ldots\ldots. a_1 A^2 + a_0 A = 0$$

$$\Rightarrow \quad A^{n+1} = \frac{-1}{a_n}(a_{n-1}A^n + a_{n-2}A^{n-1} + \ldots\ldots\ldots a_0 A) \qquad ----P219$$

Formula $P219$ allows us to know the power of a matrix, knowing its preceding powers. $P219$ is not practical because it requires us to obtain all the preceding positive power of A^k in order to determine A^k. This is too demanding.

A more powerful approach is to take advantage of the fact that

$\lambda^k = Q(\lambda)f(\lambda) + R(\lambda)$, the remainder theorem of polynomial, where λ is the eigenvalue of an nxn matrix A. λ^k is itself a polynomial expression. If we divide λ^k by the characteristic polynomial $f(\lambda)$, which is of degree n, then the remainder $R(\lambda)$ must be of degree at most $(n-1)$. That is

$$\lambda^k = Q(\lambda)f(\lambda) + R(\lambda)$$

234

Because λ is an eigenvalue of $A(nxn)$ whose characteristic polynomial is $f(\lambda)$, Calay Hamilton's theorem tells us that matrix A satisfies the equation.

$$\Rightarrow f(\lambda) = 0$$

$$\Rightarrow f(A) = 0$$

$$\therefore \quad \lambda^k = Q(\lambda)f(\lambda) + R(\lambda)$$

$$\Rightarrow A^k = (Q(A)f(A) + R(A))I$$

$$\Rightarrow A^k = I(R(A) , \; since \; f(A) = 0$$

Where $R(A)$ is a polynomial in A of degree at most $(n-1)$

$$A^k = I\big(R(A)\big) \qquad\qquad ----- P220$$

Example 5.12

Let $A = \begin{pmatrix} 1 & 4 \\ 2 & 3 \end{pmatrix}$

Find A^{-1} and A^{50}

a) By appealing to the Caley Hamilton's theorem and reduced polynomial.

b) By Diagonalizing A.

Solution

a) We first find the characteristics polynomial of A. Let λ be an eigenvalue of A, then

$$|\lambda I - A| = 0$$

$$\Rightarrow \begin{vmatrix} \lambda - 1 & -4 \\ -2 & \lambda - 3 \end{vmatrix} = 0$$

$$\Rightarrow (\lambda - 1)(\lambda - 3) - 8 = 0$$

$$\Rightarrow \lambda^2 - 4\lambda - 5 = 0$$

$$\Rightarrow \lambda = 5 , -1$$

Caley-Hamilton's theorem states that

$$A^2 - 4A - 5I = 0 \text{ ; verify this.}$$

Multiplying both sides by A^{-1}, we have

$$A^{-1}A^2 - 4A^{-1} - 5A^{-1} = 0$$

$$\Rightarrow 5A^{-1} = A - 4I$$

$$\Rightarrow A^{-1} = \frac{A - 4I}{5}$$

$$\Rightarrow A^{-1} = \frac{1}{5}\left[\begin{pmatrix} 1 & 4 \\ 2 & 3 \end{pmatrix} - \begin{pmatrix} 4 & 0 \\ 0 & 4 \end{pmatrix}\right]$$

$$\Rightarrow A^{-1} = \frac{1}{5}\begin{pmatrix} -3 & 4 \\ 2 & -1 \end{pmatrix}$$

$$= -\frac{1}{5}\begin{pmatrix} 3 & -4 \\ -2 & 1 \end{pmatrix}, \text{ as expected from the usual method.}$$

To calculate A^{50}, we appeal to the fact that λ, the eigenvalue of A, is such that $\lambda^{50} = q\lambda + r$ for some scalars q, r. This is because we already know that dividing λ^{50} by its characteristic polynomial, which is of degree 2, leaves a reminder polynomial of at most degree one. That is

$$\lambda^{50} = Q(\lambda)|\lambda I - A| + R(\lambda)$$

Since $|\lambda I - A| = 0$

$$\Rightarrow \lambda^{50} = R(\lambda) = q\lambda + r$$

Of course, not any real λ can satisfy this last equation because it takes substituting the eigenvalues of A in order to reduce the polynomial λ^{50} to a linear polynomial; so only the eigenvalues of A satisfy the equation

$$\lambda^{50} = q\lambda + r$$

When $\lambda = -1$, we have

$$(-1)^{50} = -q + r$$

$$\Rightarrow 1 = -q + r$$

When $\lambda = 5$, we have

$$5^{50} = 5q + r$$

Solving simultaneously, we have

$$6q = 5^{50} - 1$$

$$\Rightarrow q = \frac{5^{50} - 1}{6}$$

$$6r = 5^{50} + 5$$

$$r = \frac{5^{50} + 5}{6}$$

$$\lambda^{50} = q\lambda + r$$

$$A^{50} = qA + rI$$

$$A^{50} = \frac{5^{50} - 1}{6} \begin{pmatrix} 1 & 4 \\ 2 & 3 \end{pmatrix} + \frac{5^{50} + 5}{6} \begin{pmatrix} 1 & 0 \\ 0 & 1 \end{pmatrix}$$

$$A^{50} = \frac{1}{3} \begin{pmatrix} 5^{50} + 2 & 2(5^{50} - 1) \\ 5^{50} - 1 & 2 \times 5^{50} + 1 \end{pmatrix}$$

To diagonalize A, we need to further find the eigenvectors of A

For $\lambda = 5$, we have $\begin{pmatrix} 4 & -4 \\ -2 & 2 \end{pmatrix} \begin{pmatrix} x_1 \\ x_2 \end{pmatrix} = \begin{pmatrix} 0 \\ 0 \end{pmatrix}$

$$\Rightarrow x_1 = x_2 = t \; ; \; t \text{ is a scalar.}$$

$\therefore$ the eigenvector corresponding to the eigenvalue $\lambda = 5$ is

$$x = \begin{pmatrix} 1 \\ 1 \end{pmatrix}$$

For $\lambda = -1$, we have

$$\begin{pmatrix} -2 & -4 \\ -2 & -4 \end{pmatrix} \begin{pmatrix} x_1 \\ x_2 \end{pmatrix} = \begin{pmatrix} 0 \\ 0 \end{pmatrix}$$

$$\Rightarrow 2x_1 = -4x_2 \; ; \; = x_1 = -2x_2$$

$$\Rightarrow x = \begin{pmatrix} -2 \\ 1 \end{pmatrix} \text{ or } \begin{pmatrix} 2 \\ -1 \end{pmatrix}$$

$$D = P^{-1}AP$$

$$\Rightarrow A = PDP^{-1}$$

Where

$$D = \begin{pmatrix} 5 & 0 \\ 0 & -1 \end{pmatrix}, \qquad P = \begin{pmatrix} 1 & 2 \\ 1 & -1 \end{pmatrix}$$

$$\Rightarrow \begin{pmatrix} 1 & 4 \\ 2 & 3 \end{pmatrix} = \begin{pmatrix} 1 & 2 \\ 1 & -1 \end{pmatrix} \begin{pmatrix} 5 & 0 \\ 0 & -1 \end{pmatrix} \begin{pmatrix} 1 & 2 \\ 1 & -1 \end{pmatrix}^{-1}$$

$$\Rightarrow \begin{pmatrix} 1 & 4 \\ 2 & 3 \end{pmatrix}^{50} = \left[\begin{pmatrix} 1 & 2 \\ 1 & -1 \end{pmatrix} \begin{pmatrix} 5 & 0 \\ 0 & -1 \end{pmatrix} \begin{pmatrix} 1 & 2 \\ 1 & -1 \end{pmatrix} \right]^{50}$$

From what we have already established

$$\begin{pmatrix} 1 & 4 \\ 2 & 3 \end{pmatrix}^{50} = \begin{pmatrix} 1 & 2 \\ 1 & -1 \end{pmatrix} \begin{pmatrix} 5 & 0 \\ 0 & -1 \end{pmatrix}^{50} \begin{pmatrix} 1 & 2 \\ 1 & -1 \end{pmatrix}^{-1}$$

$$\Rightarrow A^{50} = \begin{pmatrix} 1 & 2 \\ 1 & -1 \end{pmatrix} \begin{pmatrix} 5^{50} & 0 \\ 0 & 1 \end{pmatrix} \begin{pmatrix} 1 & 2 \\ 1 & -1 \end{pmatrix}^{-1}$$

$$= \frac{1}{3} \begin{pmatrix} 1 & 2 \\ 1 & -1 \end{pmatrix} \begin{pmatrix} 5^{50} & 0 \\ 0 & 1 \end{pmatrix} \begin{pmatrix} 1 & 2 \\ 1 & -1 \end{pmatrix}$$

$$= \frac{1}{3} \begin{pmatrix} 5^{50} & 2 \\ 5^{50} & -1 \end{pmatrix} \begin{pmatrix} 1 & 2 \\ 1 & -1 \end{pmatrix}$$

$$= \frac{1}{3} \begin{pmatrix} 5^{50} + 2 & 2 \times 5^{50} - 2 \\ 5^{50} - 1 & 2 \times 5^{50} + 1 \end{pmatrix}$$

$$\Rightarrow A^{50} = \frac{1}{3} \begin{pmatrix} 5^{50} + 2 & 2(5^{50} - 1) \\ 5^{50} - 1 & 2 \times 5^{50} + 1 \end{pmatrix}$$

Note: in the case of the reduced polynomial

$$R(\lambda) = q\lambda + r$$

If λ is a repeated root, we simply differentiate $R(\lambda)$ wrt λ and then resubstitute λ; that is, you consider both $R(\lambda)$ and $R'(\lambda)$ to obtain q and r.

In general, if λ is a root of $|\lambda I - A|$ with multiplicity k, then we need to differentiate $R(\lambda)$ $(k-1)$ times in order to have just enough equations to obtain all the constant coefficients in $R(\lambda)$.

Let A be an nxn matrix. Then distinct eigenvectors of A corresponding to distinct eigenvalues of A are linearly independent.

Proof

Let the distinct eigenvalues of A be $\lambda_1, \lambda_2, \ldots\ldots\ldots\lambda_m$; $m \leq n$, and $\lambda_i \neq \lambda_j$ for every $i \neq j$. Let $\boldsymbol{p}_1, \boldsymbol{p}_2, \ldots\ldots\ldots\boldsymbol{p}_m$ be the eigenvector of A corresponding, respectively, to each of these eigenvalues, then

$$A\boldsymbol{p}_i = \lambda_i\boldsymbol{p}_i \text{ for each } i = 1, 2, \ldots m.$$

This, of course, does not mean that the set $\{\boldsymbol{p}_1, \boldsymbol{p}_2, \ldots\ldots\ldots\boldsymbol{p}_m\}$ exhausts all the eigenvectors of A since two of more eigenvectors can correspond to the same eigenvalue. However, this does not lead to any loss in the generality of our proof.

Since $\boldsymbol{p}_1 \neq \boldsymbol{0}$, the set $\{\boldsymbol{p}_1\}$ is linearly independent.

Let $k \leq m$ be the largest integer such that the set $\{\boldsymbol{p}_1, \boldsymbol{p}_2, \ldots\ldots\ldots\boldsymbol{p}_m\}$ is linearly independent.

Now if $k \neq m$, then $k < m$, and thus the set

$\{\boldsymbol{p}_1, \boldsymbol{p}_2, \ldots\ldots\ldots\boldsymbol{p}_{k+1}\}$ is linear dependent. Then

$$a_1\boldsymbol{p}_1 + a_2\boldsymbol{p}_2 + \ldots\ldots a_{k+1}\boldsymbol{p}_{k+1} = \boldsymbol{0} \qquad \ldots\ldots (a)$$

$\Rightarrow$ Not all a_i are zero for $i = 1, 2 \ldots\ldots k + 1$

Then it must be true that

$$A(a_1\boldsymbol{p}_1 + a_2\boldsymbol{p}_2 + \ldots\ldots\ldots a_{k+1}\boldsymbol{p}_{k+1}) = A\boldsymbol{0}$$

$$\Rightarrow a_1 A\boldsymbol{p}_1 + a_2 A\boldsymbol{p}_2 + \ldots\ldots\ldots a_{k+1} A\boldsymbol{p}_{k+1} = \boldsymbol{0}$$

$$\Rightarrow \qquad a_1\lambda_1\boldsymbol{p}_1 + a_2\lambda_2\boldsymbol{p}_2 + \ldots\ldots\ldots a_{k+1}\lambda_{k+1}\boldsymbol{p}_{k+1} = \boldsymbol{0} \qquad \ldots\ldots (b)$$

Multiplying (a) by λ_{k+1}, we have

$$a_1\lambda_{k+1}\boldsymbol{p}_1 + a_2\lambda_{k+1}\boldsymbol{p}_2 + \ldots\ldots\ldots\ldots\ldots\ldots a_{k+1}\lambda_{k+1}\boldsymbol{p}_{k+1} = 0 \qquad \ldots\ldots (c)$$

Subtracting (c) from (b), we have

$$(\lambda_1 - \lambda_{k+1})a_1 \boldsymbol{p}_1 + (\lambda_2 - \lambda_{k+1})a_2 \boldsymbol{p}_2 + \dots\dots\dots\dots (\lambda_k - \lambda_{k+1})a_k \boldsymbol{p}_k = \boldsymbol{0}$$

Since $\boldsymbol{p}_1, \boldsymbol{p}_2, \dots \dots \boldsymbol{p}_k$ are linearly independent, then

$$(\lambda_i - \lambda_{k+1})a_i = 0 \text{ for each } i = 1, 2, \dots \dots k$$

Since $\lambda_i \neq \lambda_{k+1}$,

$$\Rightarrow a_i = 0 \text{ for each } i = 1, 2, \dots \dots k$$

This, in turn, implies that

$$a_{k+1} \boldsymbol{p}_{k+1} = \boldsymbol{0}$$

Since $\boldsymbol{p}_{k+1} \neq \boldsymbol{0}$

$$\Rightarrow a_{k+1} = 0$$

Therefore, by induction, $k = m$. Thus, the eigenvectors $\boldsymbol{p}_1, \boldsymbol{p}_2 \dots \boldsymbol{p}_n$ are linearly independent.

- An immediate consequence of the above fact is that it is sufficient (but not necessary) that an $n x n$ matrix has n distinct eigenvalues to be similar to a diagonal matrix (diagonalizable).

SYMMETRIC MATRICES: SPECTRAL THEOREM

It is worth giving special attention to symmetric matrices because of their frequent occurrence in linear models, and the fact that they exhibit some special properties. Some of these have been stated before.

The Spectral Theorem

- Let A be any real symmetric matrix, then there exist a real nonsingular matrix P such that $P^{-1}AP = D$ is a diagonal matrix. In other words, every real symmetric matrix is diagonalizable; every real symmetric matrix is similar to some diagonal matrix. This is called the *spectral theorem*.

We have encountered this property before in our previous examples. The spectral theorem also tells us that every real symmetric matrix has only real eigenvalues.

- *If A is any symmetric matrix, then eigenvectors corresponding to distinct eigenvalues are orthogonal.*

Proof

Let λ_1 and λ_2 be distinct eigenvalues for any symmetric matrix A and let $\boldsymbol{p}_1$ and $\boldsymbol{p}_2$ be their corresponding eigenvectors considered as nx1 matrices (column vectors), then the matrix

$\boldsymbol{p}_1{}^T A \, \boldsymbol{p}_2$ is also necessarily symmetric since it is a singleton matrix, a 1x1 matrix.

$$\Rightarrow (\boldsymbol{p}_1{}^T A \, \boldsymbol{p}_2)^T = \boldsymbol{p}_1{}^T A \boldsymbol{p}_2$$

$$\Rightarrow (A\boldsymbol{p}_2)^T (\boldsymbol{p}_1{}^T)^T = \boldsymbol{p}_1{}^T A \, \boldsymbol{p}_2$$

$$\Rightarrow \boldsymbol{p}_2{}^T A^T \, \boldsymbol{p}_1 = \boldsymbol{p}_1{}^T A \, \boldsymbol{p}_2$$

$$\Rightarrow \boldsymbol{p}_2{}^T A \, \boldsymbol{p}_1 = \boldsymbol{p}_1{}^T A \, \boldsymbol{p}_2$$

But $A\boldsymbol{p}_1 = \lambda_1 \boldsymbol{p}_1$

$$A\boldsymbol{p}_2 = \lambda_2 \boldsymbol{p}_2$$

$$\Rightarrow \boldsymbol{p}_2{}^T \lambda_1 \, \boldsymbol{p}_1 = \boldsymbol{p}_1{}^T \lambda_2 \, \boldsymbol{p}_2$$

$$\Rightarrow \lambda_1 \boldsymbol{p}_2{}^T \, \boldsymbol{p}_1 = \lambda_2 \boldsymbol{p}_1{}^T \, \boldsymbol{p}_2$$

But $\boldsymbol{p}_2{}^T \, \boldsymbol{p}_1$ is symmetric, being a 1x1 matrix

$$\Rightarrow \boldsymbol{p}_2{}^T \, \boldsymbol{p}_1 = (\boldsymbol{p}_2{}^T \, \boldsymbol{p}_1)^T$$

$$\Rightarrow \boldsymbol{p}_2{}^T \, \boldsymbol{p}_1 = \boldsymbol{p}_1{}^T \, \boldsymbol{p}_2$$

$$\Rightarrow \lambda_1 \boldsymbol{p}_2{}^T \, \boldsymbol{p}_1 = \lambda_2 \boldsymbol{p}_2{}^T \, \boldsymbol{p}_1$$

$$\Rightarrow (\lambda_1 - \lambda_2) \boldsymbol{p}_2{}^T \, \boldsymbol{p}_1 = 0$$

But

$$\lambda_1 \neq \lambda_2$$

$$\Rightarrow \; p_2{}^T p_1 = 0$$

$$\Rightarrow \langle p_2, p_1 \rangle = \langle p_1, p_2 \rangle = 0, \; \text{proving the theorem.}$$

You can confirm from our previous examples that eigenvectors corresponding to distinct eigenvalues of a symmetric matrix are mutually orthogonal. This compels us to state the following.

- *Every symmetric matrix is orthogonally diagonalizable – meaning that if A is symmetric, then there is always an orthogonal matrix P such that*

  $D = P^{-1}AP$ *is diagonal.*

Recall that a matrix A is said to be orthogonal if and only if $A^T = A^{-1}$ - meaning that $A^T A = A^{-1}A = I$

Equivalently, a matrix A is said to be orthogonal if and only if its column vectors (or row vectors) are mutually (pairwise) orthonormal.

Therefore, the fact that the eigenvectors of a symmetric matrix A are pairwise orthogonal means that we can easily construct a matrix P for A that orthogonally diagonalizes A. This is achieved by

1. Normalizing each of the eigenvectors of A (the columns of P) if they all correspond to different eigenvalues.

2. Applying the Gram-Schmidt process of orthonormalization to eigenvectors of the same eigenvalues. We show this with the following examples.

Example 5.13

Find the matrix P that orthogonally diagonalizes the matrix

$$A = \begin{pmatrix} 1 & 1 & 1 \\ 1 & 1 & 1 \\ 1 & 1 & 1 \end{pmatrix}$$

Solution

Let λ be an eigenvalue of A, then

$$|\lambda I - A| = 0$$

$$\Rightarrow \begin{vmatrix} \lambda - 1 & -1 & -1 \\ -1 & \lambda - 1 & -1 \\ -1 & -1 & \lambda - 1 \end{vmatrix} = 0$$

Expanding along the first row, we have

$$(\lambda - 1)((\lambda - 1)^2 - 1) - (\lambda - 1) - 1 - (1 + \lambda - 1) = 0$$

$$\Rightarrow (\lambda - 1)(\lambda - 2)\lambda - 2\lambda = 0$$

$$\Rightarrow \lambda^2 (\lambda - 3) = 0$$

$$\Rightarrow \lambda = 0, repeated, \lambda = 3$$

We, therefore, expect to have an eigenspace of dimension two corresponding to $\lambda = 0$.

For $\lambda = 0$, we have

$$\begin{pmatrix} -1 & -1 & -1 \\ -1 & -1 & -1 \\ -1 & -1 & -1 \end{pmatrix} \begin{pmatrix} x_1 \\ x_2 \\ x_3 \end{pmatrix} = \begin{pmatrix} 0 \\ 0 \\ 0 \end{pmatrix}$$

$$\begin{pmatrix} -1 & -1 & -1 & | & 0 \\ -1 & -1 & -1 & | & 0 \\ -1 & -1 & -1 & | & 0 \end{pmatrix} (R_3 - R_1)(R_2 - R_1) \begin{pmatrix} -1 & -1 & -1 & | & 0 \\ 0 & 0 & 0 & | & 0 \\ 0 & 0 & 0 & | & 0 \end{pmatrix}$$

$\Rightarrow x_1 + x_2 + x_3 = 0$, an equation of a plane. This is obviously of dimension 2.

$$\Rightarrow x = \begin{pmatrix} x_1 \\ x_2 \\ x_3 \end{pmatrix} = \begin{pmatrix} -x_2 - x_3 \\ x_2 \\ x_3 \end{pmatrix} = -x_2 \begin{pmatrix} 1 \\ -1 \\ 0 \end{pmatrix} - x_3 \begin{pmatrix} 1 \\ 0 \\ -1 \end{pmatrix}$$

For any real scalars x_2 *and* x_3.

Therefore the eigenvectors corresponding to the eigenvalue $\lambda = 0$ are

$$x = \begin{pmatrix} 1 \\ -1 \\ 0 \end{pmatrix} \text{ and } \begin{pmatrix} 1 \\ 0 \\ -1 \end{pmatrix}$$

For $\lambda = 3$, we have

$$\begin{pmatrix} 2 & -1 & -1 \\ -1 & 2 & -1 \\ -1 & -1 & 2 \end{pmatrix} \begin{pmatrix} x_1 \\ x_2 \\ x_3 \end{pmatrix} = \begin{pmatrix} 0 \\ 0 \\ 0 \end{pmatrix}$$

$$\left(\begin{array}{ccc|c} 2 & -1 & -1 & 0 \\ -1 & 2 & -1 & 0 \\ -1 & -1 & 2 & 0 \end{array}\right) (R_1 + (R_2 + R_3)) \left(\begin{array}{ccc|c} 0 & 0 & 0 & 0 \\ -1 & 2 & -1 & 0 \\ -1 & -1 & 2 & 0 \end{array}\right)$$

$$(R_1 \tilde{\ } R_3), (R_2 - R_1) \left(\begin{array}{ccc|c} -1 & -1 & 2 & 0 \\ 0 & 3 & -3 & 0 \\ 0 & 0 & 0 & 0 \end{array}\right)$$

$\Rightarrow x_2 = x_3, -x_1 - x_2 = s$, for all real scalars s

$$\Rightarrow x = \begin{pmatrix} x_1 \\ x_2 \\ x_3 \end{pmatrix} = \begin{pmatrix} s \\ s \\ s \end{pmatrix} = s \begin{pmatrix} 1 \\ 1 \\ 1 \end{pmatrix}$$

Therefore the eigenvector corresponding to the eigenvalue $\lambda = 3$ is

$$x = \begin{pmatrix} 1 \\ 1 \\ 1 \end{pmatrix}$$

Now to find P that orthogonally diagonalizes A, we need column vectors (and hence row vectors) that are pairwise orthonormal. We have established already that because A is symmetric, the eigenvector $x = \begin{pmatrix} 1 \\ 1 \\ 1 \end{pmatrix}$ must be orthogonal to each of $\begin{pmatrix} 1 \\ -1 \\ 0 \end{pmatrix}$ and $\begin{pmatrix} 1 \\ 0 \\ -1 \end{pmatrix}$. It is easy to verify this. However, we need to wield the Gram-Schmidt process on the vectors $\begin{pmatrix} 1 \\ -1 \\ 0 \end{pmatrix}$ and $\begin{pmatrix} 1 \\ 0 \\ -1 \end{pmatrix}$ in

order to make them orthogonal. We go.

As usual, we first normalize $\begin{pmatrix} 1 \\ -1 \\ 0 \end{pmatrix}$. There is no regard to order. It could be the

other vector. Let

$$\begin{pmatrix} 1 \\ -1 \\ 0 \end{pmatrix} = p_1$$

$$\frac{p_1}{|p_1|} = \begin{pmatrix} \frac{1}{\sqrt{2}} \\ \frac{1}{\sqrt{2}} \\ 0 \end{pmatrix} = p_1'$$

Let

$$p_2 = \begin{pmatrix} 1 \\ 0 \\ -1 \end{pmatrix}$$

$$\therefore p_2 \perp p_1' = p_2 - \langle p_2, p_1' \rangle p_1'$$

$$= \begin{pmatrix} 1 \\ 0 \\ -1 \end{pmatrix} - \langle (1,0,-1), \left(\frac{1}{\sqrt{2}}, \frac{-1}{\sqrt{2}}, 0 \right) \rangle p_1'$$

$$= \begin{pmatrix} 1 \\ 0 \\ -1 \end{pmatrix} - \frac{1}{\sqrt{2}} \begin{pmatrix} \frac{1}{\sqrt{2}} \\ \frac{-1}{\sqrt{2}} \\ 0 \end{pmatrix}$$

$$= \begin{pmatrix} \frac{1}{2} \\ \frac{1}{2} \\ -1 \end{pmatrix}$$

We then normalize to have

$$\begin{pmatrix} \dfrac{1}{\sqrt{6}} \\ \dfrac{1}{\sqrt{6}} \\ \dfrac{-2}{\sqrt{6}} \end{pmatrix} = \boldsymbol{p}_2'$$

Note also that we need to normalize the unique (isolated) eigenvector

$$\begin{pmatrix} 1 \\ 1 \\ 1 \end{pmatrix} = \boldsymbol{p}_3$$

$$\Rightarrow \boldsymbol{p}_3' = \begin{pmatrix} \dfrac{1}{\sqrt{3}} \\ \dfrac{1}{\sqrt{3}} \\ \dfrac{1}{\sqrt{3}} \end{pmatrix}$$

Finally we arrange $\boldsymbol{p}_1', \boldsymbol{p}_2', \boldsymbol{p}_3'$ in columns to form our matrix P that is orthogonally diagonalizes A.

$$\Rightarrow P = (\boldsymbol{p}_1' : \boldsymbol{p}_2' : \boldsymbol{p}_3')$$

$$P = \begin{pmatrix} \dfrac{1}{\sqrt{3}} & \dfrac{1}{\sqrt{6}} & \dfrac{1}{\sqrt{2}} \\ \dfrac{1}{\sqrt{3}} & \dfrac{1}{\sqrt{6}} & \dfrac{-1}{\sqrt{2}} \\ \dfrac{1}{\sqrt{3}} & \dfrac{-2}{\sqrt{6}} & 0 \end{pmatrix}$$

Check that P is an orthogonal matrix (pairwise orthonormality of column vectors). Therefore $D = P^{-1}AP$ is

$$D = \begin{pmatrix} 0 & 0 & 0 \\ 0 & 0 & 0 \\ 0 & 0 & 3 \end{pmatrix}$$

Note: if the eigenvectors of a symmetric matrix A belong to distinct eigenvalues there will be no need to apply Gram-Schmidt process because they will be pairwise orthogonal. All we need to do, in this case, is just to normalize each eigenvector and then approriately arrange them in colums forming the matrix P that orthogonally diagonalises A.

Let's end this section with the following facts.

- If A and B are each orthogonal matrices, then AB is also an orthogonal matrix.

Proof

A, B are orthogonal implies

$$A^T = A^{-1}, \qquad B^T = B^{-1}$$
$$\Rightarrow \qquad (AB)^T = B^T A^T$$
$$= B^{-1} A^{-1}$$
$$= (AB)^{-1}, \text{ proved.}$$

It is easy to see that the matrix BA is also orthogonal. This is because no condition, apart from orthgonality, is imposed on either A or B.

- Let P be any orthogonal matrix, then

$$|P| = \pm 1.$$

Proof:

P is orthogonal

$$\Rightarrow \quad P^T = P^{-1}$$
$$\Rightarrow |P^T| = |P^{-1}|$$
$$\Rightarrow \quad |P| = \frac{1}{|P|}$$
$$\Rightarrow |P|^2 = 1$$
$$\Rightarrow \quad |P| = \pm 1$$

Check to see that all our previous orthogonal matrices satisfy this equation. Moreover, we can always make $|P| = 1$ by interchanging any two columns. We will need to carry out this in the following sections. This is because, for some purposes, it is appropriate to maintain $|P| = 1$.

Application to Quadratic Forms

Symmetric matrices naturally find a place in the study of analytical geometry. Here, we shall use them to study *qudratic forms* and their graphs – the classical curves called *conic sections* or simply conics. The most general quadratic equation in two variables is of the form

$$ax^2 + 2bxy + cy^2 + dx + ey + f = 0 \qquad ----- P221$$

Where a, b and c are not all zero. Note that the co-efficient of the mixed term (the "xy" term) has been chosen to be $2b$ just for convenience but without any loss of generality. In particular, the expression

$$ax^2 + 2bxy + cy^2$$

is called the qudratic form associated with P

For instance, the quadratic form associated with the quadratic equation

$$2x^2 - xy + y^2 + x - 3y - 1 = 0 \quad \text{is } 2x^2 - xy + y^2$$

Similarly, the qudratic form associated with

$$3xy - 7x + 3y + 4 = 0 \text{ is } 3xy$$

In general, the qudratic form of a quadratic equation in x and y is the sum of all the terms whose exponents in x and y sum up to 2. The graph of the quadratic equation

$$4x^2 + 3xy + 2x^2 - 5y,$$

for instance, is an ellipse. This, of course, cannot be determined by merely completing the square because it cannot be directly written in the form

$$\frac{(x-u)^2}{a^2} + \frac{(y-v)^2}{b^2} = 1$$, the standard equation of an ellipse.

The difficulty is due to the presence of the mixed term in the former equation; otherwise we would have simply completed the squares and then see that the equation describes an ellipse.

The natural question is: how then do we get rid of the mixed term? Actually, the mixed term serves to rotate our classical conics by some degrees from the coordinate axes. Therefore, getting rid of the "xy" term means "rotating out" the "xy" term through some angular degrees so as to easily ascertain which conic section is described by the quadratic equation. To achieve this, we now employ the use of orthogonal diagonalization of symmetric matrices which we have earlier carried out just for its own sake. Before we start our analysis, let us learn how to represent the quadratic equation in matrix form.

Example 5.14

Represent the quadratic equation

$$ax^2 + 2bxy + cy^2 + dx + ey + f = 0$$ in matrix form.

Solution

First the quadratic form is

$$ax^2 + 2bxy + cy^2$$

The symmetric A that can be obtained from this is

$$A = \begin{pmatrix} a & b \\ b & c \end{pmatrix}$$

If the quadratic form is $dx^2 + exy + fy^2$, then

$$A = \begin{pmatrix} d & \frac{e}{2} \\ \frac{e}{2} & f \end{pmatrix}$$

The symmetric matrix A is called the *matrix of the quadratic form.*

Let

$$X = \begin{pmatrix} x \\ y \end{pmatrix}, \qquad \Rightarrow X^T = (x, y)$$

Therefore, the quadratic form of the whole quadratic equation is represented as $X^T AX$.

To confirm this,

$$X^T AX$$

$$\Rightarrow (x, y) \begin{pmatrix} a & b \\ b & c \end{pmatrix} \begin{pmatrix} x \\ y \end{pmatrix}$$

$$\Rightarrow (x, y) \begin{pmatrix} ax + by \\ bx + cy \end{pmatrix}$$

$$\Rightarrow \big(x(ax + by) + y(bx + cy)\big), \text{ a singleton (1x1) matrix}$$

$$\Rightarrow ax^2 + 2bxy + cy^2$$

Henceforth, we adopt the convention that a singleton matrix (m) will not be distinguished from its single element m.

$$\therefore\ ax^2 + 2bxy + cy^2 \equiv X^T AX \qquad\qquad ----- P222$$

Now for the whole quadratic equation, we set

$N = (d, e), f = (f)$, a singleton (1x1) matrix

$\Rightarrow ax^2 + 2bxy + cy^2 + dx + ey + f = 0$ is equivalent, in matrix form, to

$$X^T AX + NX + f = 0 \qquad\qquad ----- P223$$

This is the matrix representation of the quadratic equation, and $X^T AX$ is the associated quadratic form.

Now we state the steps involved in eliminating the mixed term from our quadratic equation.

Step 1: Find a matrix P that orthogonally diagonalizes A, the symmetric matrix of the associated quadratic form.

Step 2: Interchange the columns of P, if necessary, so that $|P| = 1$, assuring us that P represents a rotation of the plane.

Step 3: Introduce a transformation of coordinates as follows:

$$X = PX', \text{ where } X' = \begin{pmatrix} x' \\ y' \end{pmatrix}$$

Step 4: Substitute $X = PX'$ into the equation

$X^T AX + NX + f = 0$, so that we have

$$(PX')^T APX' + NPX' + f = 0$$

$\Rightarrow$ $X'^T P^T APX' + NPX' + f = 0$

Since P is orthogonal , $P^T = P^{-1}$

$\Rightarrow \qquad X'^T P^{-1} APX' + NPX' + f = 0$

Recall that

$$P^{-1}AP = D = diag\,(\lambda_1, \lambda_2)$$

$$\Rightarrow X'^T DX' + NPX' + f = 0$$

$$\Rightarrow (x', y') \begin{pmatrix} \lambda_1 & 0 \\ 0 & \lambda_1 \end{pmatrix} \begin{pmatrix} x' \\ y' \end{pmatrix} + (d, e)P \begin{pmatrix} x' \\ y' \end{pmatrix} + f = 0$$

$$\Rightarrow \qquad \lambda_1 x'^2 + \lambda_2 y'^2 + d'x' + e'y' + f = 0 \qquad ----- P224$$

The mixed term is gone. We can now complete the square and then present our rotated curve in its canonical form. Let us illustrate with some examples after summarizing our points in the following fact.

- *Let the matrix form of the general quadratic equation of a conic section be*
  $X^T AX + NX + f = 0$ *and*

  Let the matrix P orthogonally diagonalize A with $|P| = 1$ *, then* $X = PX'$
  rotates the axes so that the equation of the rotated conics contains no
  "x'y'" term and is $X'^T (P^T AP)X' + NPX' + f = 0$

The proof has been demonstrated already.

Example 5.15

Identify the conic whose equation is

a) $\quad 53x^2 - 72xy + 32y^2 = 80$

b) $\quad xy = 2$

Solution

We only need to bring out the matrix of the associated quadratic form A and then find a matrix P that orthogonally diagonalizes A. Thereafter, we transform the given quadratic equation.

The symmetric matrix A of the associated quadratic form is given by

$$A = \begin{pmatrix} 53 & -36 \\ -36 & 32 \end{pmatrix}$$

Let λ be the eigenvalue of A, then

$$|\lambda I - A| = 0$$

$$\Rightarrow \begin{vmatrix} \lambda - 53 & 36 \\ 36 & \lambda - 32 \end{vmatrix} = 0$$

$$\Rightarrow \quad (\lambda - 53)(\lambda - 32) - 1296 = 0$$

$$\lambda^2 - 85\lambda + 400 = 0$$

$$\Rightarrow \lambda = 80, \lambda = 5$$

For $\lambda = 80$, we have

$$\begin{pmatrix} 27 & 36 \\ 36 & 48 \end{pmatrix} \begin{pmatrix} x_1 \\ x_2 \end{pmatrix} = \begin{pmatrix} 0 \\ 0 \end{pmatrix}$$

$$\Rightarrow 3x_1 = -4x_2$$

$$\Rightarrow x = \begin{pmatrix} x_1 \\ x_2 \end{pmatrix} = \begin{pmatrix} \frac{-4}{3}x_2 \\ x_2 \end{pmatrix} = \frac{-x_2}{3} \begin{pmatrix} 4 \\ -3 \end{pmatrix}, \text{ for all real } x_2$$

Therefore, the corresponding eigenvector for $\lambda = 80$ is

$$x = \begin{pmatrix} 4 \\ -3 \end{pmatrix}$$

for $\lambda = 5$, we have

$$\begin{pmatrix} -48 & 36 \\ 36 & -27 \end{pmatrix} \begin{pmatrix} x_1 \\ x_2 \end{pmatrix} = \begin{pmatrix} 0 \\ 0 \end{pmatrix}$$

$$\Rightarrow 4x_1 = 3x_2$$

$$\Rightarrow x = \begin{pmatrix} x_1 \\ x_2 \end{pmatrix} = \begin{pmatrix} \frac{3}{4}x_2 \\ x_2 \end{pmatrix} = \frac{1}{4}x_2 \begin{pmatrix} 3 \\ 4 \end{pmatrix}, \text{ for all real } x_2$$

Therefore, the corresponding eigenvector is

$$x = \begin{pmatrix} 3 \\ 4 \end{pmatrix}$$

Since the two eigenvectors correspond to distinct eigenvalues, they must be orthogonal as has been generally established. All we need do is normalize each eigenvector, and then assemble them in columns to form the matrix P that orthogonally diagonalizes A. Let the normalized vector x be denoted as x_N, where $x_N = \frac{x}{|x|}$

$$\Rightarrow \begin{pmatrix} 3 \\ 4 \end{pmatrix}_N = \begin{pmatrix} \frac{3}{5} \\ \frac{4}{5} \end{pmatrix}$$

$$\Rightarrow \begin{pmatrix} 4 \\ -3 \end{pmatrix}_N = \begin{pmatrix} \frac{4}{5} \\ \frac{-3}{5} \end{pmatrix}$$

$$P = \begin{pmatrix} \frac{4}{5} & \frac{3}{5} \\ \frac{-3}{5} & \frac{4}{5} \end{pmatrix}$$

$$\Rightarrow D = \begin{pmatrix} 80 & 0 \\ 0 & 5 \end{pmatrix} = \begin{pmatrix} \frac{4}{5} & \frac{3}{5} \\ \frac{-3}{5} & \frac{4}{5} \end{pmatrix}^{-1} \begin{pmatrix} 53 & -36 \\ -36 & 32 \end{pmatrix} \begin{pmatrix} \frac{4}{5} & \frac{3}{5} \\ \frac{-3}{5} & \frac{4}{5} \end{pmatrix}$$

Now we transform from X to X' such that

$$X = PX'$$

The matrix form of our quadratic equation is

$$X^T AX - 80 = 0$$

$$\Rightarrow (PX')^T APX' - 80 = 0$$

$$\Rightarrow X'^T P^T APX' - 80 = 0$$

Since P is orthogonal, $P^T AP = P^{-1}AP = D$

$$\Rightarrow X'^T DX' - 80 = 0$$

$$\Rightarrow (x', y') \begin{pmatrix} 80 & 0 \\ 0 & 5 \end{pmatrix} \begin{pmatrix} x' \\ y' \end{pmatrix} - 80 = 0$$

$$\Rightarrow 80x'^2 + 5y'^2 - 80 = 0$$

$$\Rightarrow x'^2 + \frac{y'^2}{16} = 1$$

Therefore, the conic is an ellipse rotated from the coordinate axes by angle $\emptyset$, where

$$\cos \emptyset = \frac{4}{5}, \quad \sin \emptyset = \frac{-3}{5}$$

$$\Rightarrow \emptyset = \cos^{-1}\left(\frac{4}{5}\right) = \sin^{-1}\left(\frac{-3}{5}\right)$$

Note that angle $\emptyset$ is gotten from matrix P.

$$P \equiv \begin{pmatrix} \cos\emptyset & -\sin\emptyset \\ \sin\emptyset & \cos\emptyset \end{pmatrix}$$

The major axis of our rotated ellipse is the y' – axis.

b) $xy = 2$

The associated quadratic form, in matrix form is

$$X^T A X$$

The equation in matrix form is

$$X^T A X - 2 = 0$$

where

$$A = \begin{pmatrix} 0 & \frac{1}{2} \\ \frac{1}{2} & 0 \end{pmatrix}$$

Let λ be an eigenvalue of A, we have

$$|\lambda I - A| = 0$$

$$\Rightarrow \begin{vmatrix} \lambda & \frac{1}{2} \\ \frac{1}{2} & \lambda \end{vmatrix} = 0$$

$$\Rightarrow \lambda^2 - \frac{1}{4} = 0$$

$$\Rightarrow \lambda = \frac{1}{2} , \ \lambda = \frac{-1}{2}$$

For $\lambda = \frac{1}{2}$, we have

$$\begin{pmatrix} \frac{1}{2} & \frac{1}{2} \\ \frac{1}{2} & \frac{1}{2} \end{pmatrix}\begin{pmatrix} x_1 \\ x_2 \end{pmatrix} = \begin{pmatrix} 0 \\ 0 \end{pmatrix}$$

$$\Rightarrow x_1 = -x_2 = t \text{ for all real } t$$

$\Rightarrow x = \begin{pmatrix} 1 \\ -1 \end{pmatrix}$ or $\begin{pmatrix} -1 \\ 1 \end{pmatrix}$ is the corresponding eigenvector for $\lambda = \frac{1}{2}$.

For $\lambda = \dfrac{-1}{2}$, we have

$$\begin{pmatrix} \dfrac{-1}{2} & \dfrac{1}{2} \\ \dfrac{1}{2} & \dfrac{-1}{2} \end{pmatrix} \begin{pmatrix} x_1 \\ x_2 \end{pmatrix} = \begin{pmatrix} 0 \\ 0 \end{pmatrix}$$

$$\Rightarrow x_1 = x_2 = t \text{ for all real } t$$

$x = t \begin{pmatrix} 1 \\ 1 \end{pmatrix}$. The corresponding eigenvector is $x = \begin{pmatrix} 1 \\ 1 \end{pmatrix}$.

Again, the eigenvectors $\begin{pmatrix} 1 \\ -1 \end{pmatrix}$ and $\begin{pmatrix} -1 \\ 1 \end{pmatrix}$ are orthogonal since they correspond to distinct eigenvalues. It is easily verifiable because

$$\langle (1, -1), (1, 1) \rangle = 0$$

Therefore, we need only to normalize each eigenvector, and then form our matrix P.

$$\begin{pmatrix} 1 \\ -1 \end{pmatrix}_N = \begin{pmatrix} \dfrac{1}{\sqrt{2}} \\ \dfrac{-1}{\sqrt{2}} \end{pmatrix}$$

$$\begin{pmatrix} 1 \\ 1 \end{pmatrix}_N = \begin{pmatrix} \dfrac{1}{\sqrt{2}} \\ \dfrac{1}{\sqrt{2}} \end{pmatrix}$$

$$\Rightarrow P = \begin{pmatrix} \dfrac{1}{\sqrt{2}} & \dfrac{1}{\sqrt{2}} \\ \dfrac{-1}{\sqrt{2}} & \dfrac{1}{\sqrt{2}} \end{pmatrix}$$

256

$$D = P^{-1}AP$$

$$\Rightarrow \quad D = \begin{pmatrix} \frac{1}{2} & 0 \\ 0 & \frac{-1}{2} \end{pmatrix}$$

On transformation,

$$X^T A X - 2 = 0$$

becomes

$$X'^T D X' - 2 = 0$$

$$\Rightarrow \quad \frac{1}{2}x'^2 - \frac{1}{2}y'^2 = 2$$

$$\Rightarrow \quad \frac{1}{4}x'^2 - \frac{1}{4}y'^2 = 1$$

This is obviously a hyperbola with x' - axis as the major axis. The hyperbola is rotated from the coordinate axis by $\emptyset$, where

$$\cos \emptyset = \frac{1}{\sqrt{2}} \ , \ \text{Sin } \emptyset = \frac{-1}{\sqrt{2}}$$

$$\Rightarrow \emptyset = -\frac{\pi}{4} \ or \ \emptyset = \frac{7\pi}{4}$$

Note that $|P| = 1$ for the two examples we have considered. If $|P| \neq 1$, then we interchange the columns and also interchange the diagonal elements of D.

Example 5.16

Rotate and eliminate the "xy" term from the quadratic equation $2x^2 - 3xy + 4y^2 - 7x + 2y + 7 = 0$, hence describe and sketch the graph of the conic section

Solution

As before, the symmetric matrix A of the associated quadratic form of the quadratic equation is

$$A = \begin{pmatrix} 2 & \dfrac{-3}{2} \\ \dfrac{-3}{2} & 4 \end{pmatrix}$$

The matrix form of the quadratic equation is

$$X^T A X + (-7,2)X + 7 = 0$$

Where

$$X = \begin{pmatrix} x \\ y \end{pmatrix}$$

Let λ be an eigenvalue of A, then

$$|\lambda I - A| = 0$$

$$\Rightarrow \begin{vmatrix} \lambda - 2 & \dfrac{3}{2} \\ \dfrac{3}{2} & \lambda - 4 \end{vmatrix} = 0$$

$$\Rightarrow (\lambda - 2)(\lambda - 4) - \left(\dfrac{3}{2}\right)^2 = 0$$

$$\Rightarrow 4\lambda^2 - 24\lambda + 23 = 0$$

$$\Rightarrow \lambda = \dfrac{6 + \sqrt{13}}{2}$$

For $\lambda = \dfrac{6 + \sqrt{13}}{2}$, we have

$$\begin{pmatrix} \dfrac{2 + \sqrt{13}}{2} & \dfrac{3}{2} \\ \dfrac{3}{2} & \dfrac{-2 + \sqrt{13}}{2} \end{pmatrix} \begin{pmatrix} x_1 \\ x_2 \end{pmatrix} = \begin{pmatrix} 0 \\ 0 \end{pmatrix}$$

$$\Rightarrow 3x_1 = (2 - \sqrt{13})x_2$$

$$\Rightarrow \begin{pmatrix} x_1 \\ x_2 \end{pmatrix} = \begin{pmatrix} \dfrac{2-\sqrt{13}}{3}x_2 \\ x_2 \end{pmatrix} = \dfrac{x_2}{3}\begin{pmatrix} 2 - \sqrt{13} \\ 3 \end{pmatrix} \text{ for all real } x_2$$

Therefore, the corresponding eigenvector to $\lambda = \dfrac{6 + \sqrt{13}}{2}$ is

$$x = \begin{pmatrix} 2 - \sqrt{13} \\ 3 \end{pmatrix}$$

For $\lambda = \frac{6 - \sqrt{13}}{2}$, we have

$$\begin{pmatrix} \dfrac{2 - \sqrt{13}}{2} & \dfrac{3}{2} \\ \dfrac{3}{2} & \dfrac{-2 - \sqrt{13}}{2} \end{pmatrix} \begin{pmatrix} x_1 \\ x_2 \end{pmatrix} = \begin{pmatrix} 0 \\ 0 \end{pmatrix}$$

$$\Rightarrow 3x_1 = (2 + \sqrt{13})x_2$$

$$\Rightarrow x = \begin{pmatrix} \frac{2 + \sqrt{13}}{3} x_2 \\ x_2 \end{pmatrix} = \frac{x_2}{3} \begin{pmatrix} 2 + \sqrt{13} \\ 3 \end{pmatrix}, \text{ for all real } x_2$$

Therefore, the eigenvector corresponding to $\lambda = \frac{6 - \sqrt{13}}{2}$ is

$$x = \begin{pmatrix} 2 + \sqrt{13} \\ 3 \end{pmatrix}$$

We need not test for orthogonality because the two eigenvectors correspond to distinct eigenvalues. However, it is also clear that the magnitude of each of them is not unity; thus, we need to normalize each of them.

$$\begin{pmatrix} 2 - \sqrt{13} \\ 3 \end{pmatrix}_N = \begin{pmatrix} \dfrac{2 - \sqrt{13}}{\sqrt{26 - 4\sqrt{13}}} \\ \dfrac{3}{\sqrt{26 - 4\sqrt{13}}} \end{pmatrix}$$

Similarly,

$$\begin{pmatrix} 2 + \sqrt{13} \\ 3 \end{pmatrix}_N = \begin{pmatrix} \dfrac{2 + \sqrt{13}}{\sqrt{26 + 4\sqrt{13}}} \\ \dfrac{3}{\sqrt{26 + 4\sqrt{13}}} \end{pmatrix}$$

$$\Rightarrow P = \begin{pmatrix} \dfrac{2-\sqrt{13}}{\sqrt{26-4\sqrt{13}}} & \dfrac{2+\sqrt{13}}{\sqrt{26+4\sqrt{13}}} \\[4mm] \dfrac{3}{\sqrt{26-4\sqrt{13}}} & \dfrac{3}{\sqrt{26+4\sqrt{13}}} \end{pmatrix}$$

We need to check that $|P| = 1$ as this is not obvious as the previous cases.

$$|P| = \frac{3(2-\sqrt{13})}{\sqrt{26^2 - \left(4\sqrt{13}\right)^2}} - \frac{3(2+\sqrt{13})}{\sqrt{26^2 - \left(4\sqrt{13}\right)^2}}$$

We need not continue because it is now obvious that $|P| = -1$. This is because the two terms on the RHS have the same denominator and the second term clearly has a higher numerator, this implies that $|P| = -1$ since $|P| = \pm 1$, being an orthogonal matrix as we have earlier shown. Thus, we need to interchange the columns of matrix P so that $|P| = 1$, and thus the rotation of the axes is obtained.

$$\Rightarrow P = \begin{pmatrix} \dfrac{2+\sqrt{13}}{\sqrt{26+4\sqrt{13}}} & \dfrac{2-\sqrt{13}}{\sqrt{26-4\sqrt{13}}} \\[4mm] \dfrac{3}{\sqrt{26+4\sqrt{13}}} & \dfrac{3}{\sqrt{26-4\sqrt{13}}} \end{pmatrix}$$

The direct effect of interchanging the columns of P is a change in the order of occurrence of the corresponding eigenvalues in the diagonal matrix D; the eigenvalue corresponding to the kth column of P becomes λ_{kk} in D. Therefore, the diagonal matrix D is given by

$$D = \begin{pmatrix} \dfrac{6-\sqrt{13}}{2} & 0 \\[4mm] 0 & \dfrac{6+\sqrt{13}}{2} \end{pmatrix}$$

$$X^T AX + (-7,2) X + 7 = 0$$

becomes

$$X'^T DX' + (-7,2)PX' + 7 = 0$$

$$\Rightarrow \left(\frac{6-\sqrt{13}}{2}\right)x'^2 + \left(\frac{6+\sqrt{13}}{2}\right)y'^2 + (-7,2)\,PX' + 7 = 0$$

$$(-7,2)P = \left(\left(\frac{-7(2+\sqrt{13})}{\sqrt{26+4\sqrt{13}}}\right) + \frac{6}{\sqrt{26+4\sqrt{13}}}\quad,\quad \frac{-7(2-\sqrt{13})}{\sqrt{26-4\sqrt{13}}} + \frac{6}{\sqrt{26-4\sqrt{13}}}\right)$$

$$(-7,2)P = \left(\frac{-8-7\sqrt{13}}{\sqrt{26+4\sqrt{13}}},\frac{-8+7\sqrt{13}}{\sqrt{26-4\sqrt{13}}}\right)$$

$$\Rightarrow (-7,2)PX' = \frac{-8-7\sqrt{13}}{\sqrt{26+4\sqrt{13}}}x' + \frac{-8+7\sqrt{13}}{\sqrt{26-4\sqrt{13}}}y'$$

$$\Rightarrow \left(\frac{6-\sqrt{13}}{2}\right)x'^2 + \left(\frac{6+\sqrt{13}}{2}\right)y'^2 - \left(\frac{8+7\sqrt{13}}{\sqrt{26+4\sqrt{13}}}\right)x' + \left(\frac{-8+7\sqrt{13}}{\sqrt{26-4\sqrt{13}}}\right)y' + 7 = 0$$

This is the equation of the original conic rotated from the coordinate axes

through $\emptyset = \cos^{-1}\left(\frac{2+\sqrt{13})}{\sqrt{26+4\sqrt{13}}}\right)$

Clearly, the equation describes an eclipse since the coefficients of x'^2 and y'^2 are not equal, and are of the same sign (provided it does degenerate). To write this equation in its canonical form, we need to complete the square. The interested reader may go ahead to do so.

Now we establish a more general result, the one that helps us to determine directly the graph (conic) of a quadratic equation without performing any transformation on it. Let's see.

By now, it should be very clear to us that the type of conics described by the quadratic equation

$$ax^2 + 2bxy + cy^2 + dx + ey + f = 0$$

is solely determined by the associated quadratic form

$$ax^2 + 2bxy + cy^2$$

$$A = \begin{pmatrix} a & b \\ b & c \end{pmatrix}$$

Let λ be an eigenvalue of A, then

$$|\lambda I - A| = 0$$

$$\Rightarrow \begin{vmatrix} \lambda - a & -b \\ -b & \lambda - c \end{vmatrix} = 0$$

$$\Rightarrow \quad (\lambda - a)(\lambda - c) - b^2 = 0$$

$$\lambda^2 - \lambda(a + c) + ac - b^2 = 0$$

$$\Rightarrow \quad \lambda = \frac{a + c \pm \sqrt{(a + c)^2 + 4b^2 - 4ac}}{2}$$

Let

$$q = \sqrt{(a + c)^2 + 4b^2 - 4ac}$$

$$\Rightarrow \lambda = \frac{a + c \pm q}{2}$$

For $\lambda = \frac{a + c + q}{2}$, we have

$$\begin{pmatrix} \dfrac{c + q - a}{2} & -b \\ -b & \dfrac{q + a - c}{2} \end{pmatrix} \begin{pmatrix} x_1 \\ x_2 \end{pmatrix} = \begin{pmatrix} 0 \\ 0 \end{pmatrix}$$

$$\Rightarrow (c + q - a)x_1 = 2bx_2$$

$$\Rightarrow \begin{pmatrix} x_1 \\ x_2 \end{pmatrix} = \begin{pmatrix} \dfrac{2bx_2}{c + q - a} \\ x_2 \end{pmatrix} = \frac{x_2}{c + q - a} \begin{pmatrix} 2b \\ c + q - a \end{pmatrix} ; \text{ for all real } x_2$$

Therefore, the eigenvalue corresponding to $\lambda = \frac{a + c + q}{2}$ is

$$x = \begin{pmatrix} 2b \\ c + q - a \end{pmatrix}$$

Similarly, for $\lambda = \frac{a + c - q}{2}$, we have

$$\begin{pmatrix} \dfrac{c-q-a}{2} & -b \\ -b & \dfrac{a-q-c}{2} \end{pmatrix} \begin{pmatrix} x_1 \\ x_2 \end{pmatrix} = \begin{pmatrix} 0 \\ 0 \end{pmatrix}$$

$$\Rightarrow \quad bx_1 = \left(\dfrac{a-q-c}{2}\right)x_2 \; ; \; 2bx_1 = (a-q-c)x_2$$

$$\Rightarrow \begin{pmatrix} x_1 \\ x_2 \end{pmatrix} = \begin{pmatrix} \left(\dfrac{a-q-c}{2b}\right)x_2 \\ x_2 \end{pmatrix} = \dfrac{x_2}{2b}\begin{pmatrix} a-q-c \\ 2b \end{pmatrix} \text{for all real } x_2. \text{ Therefore the}$$

corresponding eigenvector to $\lambda = \dfrac{a+c-q}{2}$ is

$$x = \begin{pmatrix} a-q-c \\ 2b \end{pmatrix}$$

$$\Rightarrow P = \begin{pmatrix} 2b & a-q-c \\ c+q-a & 2b \end{pmatrix} \text{or} \begin{pmatrix} a-q-c & 2b \\ 2b & c+q-a \end{pmatrix}$$

According as $|P| = 1 \; or -1$; we seek the one that corresponds to $|P| = 1$.

Now let us assume that $|P| = 1$, then

$$X^T A X$$

becomes

$$X'^T D X'$$

Where

$$D = \begin{pmatrix} \dfrac{a+c+q}{2} & 0 \\ 0 & \dfrac{a+c-q}{2} \end{pmatrix} \text{or} \begin{pmatrix} \dfrac{a+c-q}{2} & 0 \\ 0 & \dfrac{a+c+q}{2} \end{pmatrix}, \text{according as } P.$$

$$= \left(\dfrac{a+c+q}{2}\right)x'^2 + \left(\dfrac{a+c-q}{2}\right)y'^2$$

Or

$$\left(\dfrac{a+c-q}{2}\right)x'^2 + \left(\dfrac{a+c+q}{2}\right)y'^2$$

Which ever the case, the following facts are obvious

i)   *The number $a + c$ is invariant under any rotation – meaning that*

$$a' + c' = a + c$$

Proof

$$a' = \frac{a + c - q}{2} \quad when \quad c' = \frac{a + c + q}{2}$$

$$or$$

$$a' = \frac{a + c + q}{2} \quad when \quad c' = \frac{a + c - q}{2}$$

$$\Rightarrow \quad a' + c' = \frac{a+c+q}{2} + \frac{a+c-q}{2}$$

$$\Rightarrow \quad a' + c' = a + c \quad , \text{ hence } a + c \text{ is invariant.}$$

The number $b^2 - ac$, called the discriminat, is invariant under any rotation-meaning that $b'^2 - a'c' = b^2 - ac$

Proof

As before,

$$a' = \frac{a + c \pm q}{2} \quad , \quad c' = \frac{a + c \mp q}{2} \quad , \quad b' = 0$$

$$\Rightarrow b'^2 - a'c' = 0 - \frac{1}{4}(a + c + q)(a + c - q)$$

$$= \frac{1}{4}(q^2 - (a + c)^2)$$

But

$$q = \sqrt{(a + c)^2 + 4b^2 - 4ac}$$

$$\Rightarrow b'^2 - a'c' = \frac{1}{4}[(a + c)^2 + 4b^2 - 4ac - (a + c)^2]$$

$$= \frac{1}{4}[4b^2 - 4ac]$$

$$= b^2 - ac$$

$$\therefore b'^2 - a'c' = b^2 - ac \text{ , hence invariance.}$$

It is left to the reader to show that

$$D' = \begin{vmatrix} a & b & d \\ b & c & e \\ d & e & f \end{vmatrix} \text{ is invariant under any rotation.}$$

Now let us see how to determine the nature of a conic from its most general form of quadratic equation without transformation.

We already know that

$$b' = 0 \text{ , } \qquad a' = \frac{a + c \pm q}{2} \text{ , } \qquad c' = \frac{a + c \mp q}{2}$$

Indeed the nature of the conic is solely dependent on a' and c'.

If one of a' and c' is zero, we have a parabola. This is because

$$a' = 0 \text{ or } c' = 0$$

$$\Rightarrow a + c = q \text{ or } a + c = -q$$

$$\Rightarrow \quad |a + c| = |q|$$

$$\Rightarrow (a + c)^2 = q^2$$

$$\Rightarrow \quad (a + c)^2 = (a + c)^2 + 4b^2 - 4ac$$

$$\Rightarrow 4b^2 - 4ac = 0$$

$$\Rightarrow \quad b^2 - ac = 0$$

For an ellipse, a' and c' must have the same sign but different magnitudes.

$$\Rightarrow a' > 0 \text{ and } c' > 0$$

Or

$$a' < 0 \text{ and } c' < 0$$

Case 1:

$$a' > 0 \text{ and } c' > 0$$

$$\Rightarrow a + c + q > 0 \text{ and } a + c - q > 0$$

Since $a + c + q > a + c - q$, q being always positive for a nondegenerate case (why?).

This means that it suffices to consider only the inequality

$$a + c - q > 0$$

$$\Rightarrow (a + c)^2 > q^2 \text{ , because } q > 0$$

$$\Rightarrow (a + c)^2 > (a + c)^2 + 4b^2 - 4ac$$

$$\Rightarrow 0 > 4b^2 - 4ac$$

$$\Rightarrow b^2 - ac < 0$$

Case 2:

$$a' < 0 \text{ and } c' < 0$$

$$\Rightarrow a + c + q < 0 \text{ and } a + c - q < 0$$

Again, it suffices to consider only the inequality

$$a + c + q < 0$$

$$\Rightarrow a + c < -q$$

Since $q > 0$

$$\Rightarrow |a + c| > |q|$$

$$\Rightarrow (a + c)^2 > q^2$$

$$\Rightarrow b^2 - ac < 0$$

For a hyperbola , both a' and c' must be opposite in signs.

$$\Rightarrow a' > 0 \text{ and } c' < 0$$

$$\text{Or}$$

$$a' < 0 \text{ and } c' > 0$$

This means that the greater of a' and c' is positive, and the lesser is negative

$$\Rightarrow a + c + q > 0 \text{ and } a + c - q < 0$$

$$\Rightarrow |q| > |a + c|$$

$$\Rightarrow q^2 > (a+c)^2$$

$$\Rightarrow (a+c)^2 + 4b^2 - 4ac > (a+c)^2$$

$$\Rightarrow \quad 4b^2 - 4ac > 0$$

$$\Rightarrow \quad b^2 - ac > 0$$

In summary,

$$\left. \begin{array}{l} \textit{For an ellipse}: b^2 - ac\ < 0 \\ \textit{For a parabola}\ :\ b^2 - ac\ =\ 0 \\ \textit{For a hyperbola}:\ b^2 - ac > 0 \end{array} \right\} \qquad ----- P225$$

For example, the equation

$$xy\ =\ k\ ,\ k \text{ is constant}$$

has

$$a = c = 0\ , b\ =\ \tfrac{1}{2}$$

$$D = b^2 - ac \text{ , the discriminant}$$

$$\Rightarrow D = \left(\frac{1}{2}\right)^2 - 0\ =\ \frac{1}{4} > 0$$

$\Rightarrow$ The equation describes a hyperbola.

Note that for any of the degenerate cases, the determinant

$$D' = \begin{vmatrix} a & b & d \\ b & c & e \\ d & e & f \end{vmatrix} = 0$$

The degenerate case could be an *imaginary conic* such as $2x^2 + y^2 + 3 = 0$, which is impossible in the reals.

Or

A pair of lines , such as $x^2 - y^2 = 0$

QUADRATIC SURFACES

The graphs of quadratic forms in three variables are called *quadratic surfaces* or simply quadratics.

The most general form of quadratic equation in three variables is

$$a_{11}x^2 + a_{22}y^2 + a_{33}z^2 + 2a_{12}xy + 2a_{13}xz + 2a_{23}yz + bx + cy + dz + f = 0$$

Where not all $a_{ij} = 0$.

The associated quadratic form of this general equation is

$$a_{11}x^2 + a_{22}y^2 + a_{33}z^2 + 2a_{12}xy + 2a_{13}xz + 2a_{23}yz$$

The matrix of the quadratic form is

$$A = \begin{pmatrix} a_{11} & a_{12} & a_{13} \\ a_{21} & a_{22} & a_{23} \\ a_{31} & a_{32} & a_{33} \end{pmatrix}$$

It is not difficult to comprehend that the matrix form of the quadratic equation is $X^T A X + N X + f = 0$

Where

$$A = \begin{pmatrix} a_{11} & a_{12} & a_{13} \\ a_{21} & a_{22} & a_{23} \\ a_{31} & a_{32} & a_{33} \end{pmatrix} , \quad N = (b, c, d) , \quad X = \begin{pmatrix} x \\ y \\ z \end{pmatrix}$$

Standard Quadratics

i) Ellipsoid

$$\frac{x^2}{a^2} + \frac{y^2}{b^2} + \frac{z^2}{c^2} = 1$$

ii) Elliptic hyperboloid of one sheet

$$\frac{x^2}{a^2} + \frac{y^2}{b^2} - \frac{z^2}{c^2} = 1$$

iii)   Elliptic hyperboloid of two sheets

$$\frac{x^2}{a^2} - \frac{y^2}{b^2} - \frac{z^2}{c^2} = 1$$

iv)   Elliptic cone

$$\frac{x^2}{a^2} + \frac{y^2}{b^2} = \frac{z^2}{c^2}$$

v)   Elliptic Paraboloid

$$\frac{x^2}{a^2} + \frac{y^2}{b^2} = z$$

vi)   Hyperbolic Paraboloid

$$\frac{x^2}{a^2} - \frac{y^2}{b^2} = z$$

Like the case of two variables, the presence of the mixed terms causes a rotation of the quadratic from the coordinate axes. To determine the particular quadratic, we need to eliminate the mixed terms by rotating the axes. This is achieved, as usual, by transforming the coordinate system – defining a change of variables.

Example 5.17

Name the quadratic surface whose matrix equation is

$$(x, y, z) \begin{pmatrix} 2 & 0 & 0 \\ 0 & -1 & 0 \\ 0 & 0 & 7 \end{pmatrix} \begin{pmatrix} x \\ y \\ z \end{pmatrix} = 11$$

Solution

We need only to multiply the matrices

$$(x, y, z) \begin{pmatrix} 2 & 0 & 0 \\ 0 & -1 & 0 \\ 0 & 0 & 7 \end{pmatrix} \begin{pmatrix} x \\ y \\ z \end{pmatrix} = (x, y, z) \begin{pmatrix} 2x \\ -y \\ 7z \end{pmatrix}$$

$$2x^2 - y^2 + 7z^2 = 11$$

$$\Rightarrow \quad \frac{x^2}{11/2} - \frac{y^2}{11} + \frac{z^2}{11/7} = 1$$

This is an elliptic hyperboloid of one sheet oriented along the y-axis

Example 5.18

Identify the quadratic whose equation is

a) $\quad xy + z = 0$

b) $\quad 2xy + 2xz + 2yz - 6x - 6y - 4z = -9$

Solution

$$xy + z$$

The associated quadratic form is

$$xy$$

The matrix of the associated quadratic form is

$$A = \begin{pmatrix} 0 & \frac{1}{2} & 0 \\ \frac{1}{2} & 0 & 0 \\ 0 & 0 & 0 \end{pmatrix}$$

Let λ be an eigenvalue of A, then

$$|\lambda I - A| = 0$$

$$\Rightarrow \begin{vmatrix} \lambda & \frac{-1}{2} & 0 \\ \frac{-1}{2} & \lambda & 0 \\ 0 & 0 & \lambda \end{vmatrix} = 0$$

Expanding along the third row or third column, we have

$$\lambda \left(\lambda^2 - \frac{1}{4} \right) = 0$$

$$\Rightarrow \lambda = 0, \frac{1}{2}, \frac{-1}{2}$$

For $\lambda = 0$

$$\begin{pmatrix} 0 & \frac{-1}{2} & 0 \\ \frac{-1}{2} & 0 & 0 \\ 0 & 0 & 0 \end{pmatrix} \begin{pmatrix} x \\ y \\ z \end{pmatrix} = \begin{pmatrix} 0 \\ 0 \\ 0 \end{pmatrix}$$

$$\Rightarrow x_3 = s, x_1 = x_2 = 0$$

$$x = \begin{pmatrix} 0 \\ 0 \\ 1 \end{pmatrix}$$ is the corresponding eigenvector

For $\lambda = \frac{1}{2}$

$$\begin{pmatrix} \frac{1}{2} & \frac{-1}{2} & 0 \\ \frac{-1}{2} & \frac{1}{2} & 0 \\ 0 & 0 & \frac{1}{2} \end{pmatrix} \begin{pmatrix} x \\ y \\ z \end{pmatrix} = \begin{pmatrix} 0 \\ 0 \\ 0 \end{pmatrix}$$

$$\Rightarrow x_3 = 0, x_1 = x_2$$

$$x = \begin{pmatrix} 1 \\ 1 \\ 0 \end{pmatrix}$$ is the corresponding eigenvector

For $\lambda = -\frac{1}{2}$

$$\begin{pmatrix} \frac{-1}{2} & \frac{-1}{2} & 0 \\ \frac{-1}{2} & \frac{-1}{2} & 0 \\ 0 & 0 & \frac{-1}{2} \end{pmatrix} \begin{pmatrix} x \\ y \\ z \end{pmatrix} = \begin{pmatrix} 0 \\ 0 \\ 0 \end{pmatrix}$$

$$\Rightarrow x_3 = 0, x_1 = -x_2$$

$$\Rightarrow x = \begin{pmatrix} 1 \\ -1 \\ 0 \end{pmatrix} \text{ is the corresponding eigenvector}$$

$$\Rightarrow D = P^{-1}AP$$

$$D = \begin{pmatrix} 0 & 0 & 0 \\ 0 & \dfrac{1}{2} & 0 \\ 0 & 0 & \dfrac{-1}{2} \end{pmatrix}$$

Since the three eigenvectors corresponds to distinct eigenvalues, they must be pairwise orthogonal, and so we need not apply the Gram-Schmidt process. We only need to normalize each of them.

$$x = \begin{pmatrix} 0 \\ 0 \\ 1 \end{pmatrix} \text{ is already in normal form.}$$

$$\begin{pmatrix} 1 \\ 1 \\ 0 \end{pmatrix}_N = \begin{pmatrix} \dfrac{1}{\sqrt{2}} \\ \dfrac{1}{\sqrt{2}} \\ 0 \end{pmatrix}$$

$$\begin{pmatrix} 1 \\ -1 \\ 0 \end{pmatrix}_N = \begin{pmatrix} \dfrac{1}{\sqrt{2}} \\ \dfrac{-1}{\sqrt{2}} \\ 0 \end{pmatrix}$$

$$\Rightarrow P = \begin{pmatrix} 0 & \dfrac{1}{\sqrt{2}} & \dfrac{1}{\sqrt{2}} \\ 0 & \dfrac{1}{\sqrt{2}} & \dfrac{-1}{\sqrt{2}} \\ 1 & 0 & 0 \end{pmatrix}; |P| = -1$$

The next thing is to check whether $|P| = 1$. Obviously $|P| = -1$, and so we need to interchange any two columns. Let us interchange columns 1 and 2.

$$P = \begin{pmatrix} \frac{1}{\sqrt{2}} & 0 & \frac{1}{\sqrt{2}} \\ \frac{1}{\sqrt{2}} & 0 & \frac{-1}{\sqrt{2}} \\ 0 & 1 & 0 \end{pmatrix} ; \ |P| = 1$$

This means that our diagonal matrix D changes appropriately. Therefore D becomes

$$D = \begin{pmatrix} \frac{1}{2} & 0 & 0 \\ 0 & 0 & 0 \\ 0 & 0 & \frac{-1}{2} \end{pmatrix}$$

The quadratic form, in matrix form, is

$$X^T A X + N X = 0$$

Where

$$N = (0,0,1), \qquad X = \begin{pmatrix} x \\ y \\ z \end{pmatrix}$$

Imposing the change of coordinate axes

$$X = PX'$$

we have

$$X'^T D X' + N P X' = 0$$

$$\Rightarrow \quad (x',y',z')\begin{pmatrix} \frac{1}{2} & 0 & 0 \\ 0 & 0 & 0 \\ 0 & 0 & \frac{-1}{2} \end{pmatrix}\begin{pmatrix} x' \\ y' \\ z' \end{pmatrix} + (0,0,1)\begin{pmatrix} \frac{1}{\sqrt{2}} & 0 & \frac{1}{\sqrt{2}} \\ \frac{1}{\sqrt{2}} & 0 & \frac{-1}{\sqrt{2}} \\ 0 & 1 & 0 \end{pmatrix}\begin{pmatrix} x' \\ y' \\ z' \end{pmatrix} = 0$$

$$\Rightarrow \quad \frac{1}{2}x'^2 - \frac{1}{2}z'^2 + (0,1,0)\begin{pmatrix} x' \\ y' \\ z' \end{pmatrix} = 0$$

$$\Rightarrow \quad \frac{1}{2}x'^2 - \frac{1}{2}z'^2 + y' = 0$$

$$\Rightarrow \qquad \frac{1}{2}z'^2 - \frac{1}{2}x'^2 = y'$$

This is a hyperbolic paraboloid oriented along the y' – axis.

b) $\qquad 2xy + 2xz + 2yz - 6x - 6y - 4z = -9$

$$X^T AX + NX + 9 = 0$$

$$X = \begin{pmatrix} x \\ y \\ z \end{pmatrix}, N = (-6, -6, -4)$$

$$A = \begin{pmatrix} 0 & 1 & 1 \\ 1 & 0 & 1 \\ 1 & 1 & 0 \end{pmatrix}$$

Let λ be an eigenvalue of A, we have

$$|\lambda I - A| = 0$$

$$\Rightarrow \begin{vmatrix} \lambda & -1 & -1 \\ -1 & \lambda & -1 \\ -1 & -1 & \lambda \end{vmatrix} = 0$$

Expanding along the first row, we have

$$\lambda(\lambda^2 - 1) + -\lambda - 1(1 + \lambda) = 0$$

$$\lambda(\lambda + 1)(\lambda - 1) - 2(\lambda + 1) = 0$$

$$(\lambda + 1)(\lambda - 2)(\lambda + 1)$$

$$\Rightarrow \lambda = -1, \text{ repeated}, \lambda = 2$$

For $\lambda = -1$

$$\begin{pmatrix} -1 & -1 & -1 \\ -1 & -1 & -1 \\ -1 & -1 & -1 \end{pmatrix} \begin{pmatrix} x \\ y \\ z \end{pmatrix} = \begin{pmatrix} 0 \\ 0 \\ 0 \end{pmatrix}$$

$$\left(\begin{array}{ccc|c} -1 & -1 & -1 & 0 \\ -1 & -1 & -1 & 0 \\ -1 & -1 & -1 & 0 \end{array} \right) (R_2 - R_1), (R_3 - R_1) \underset{\sim}{} \left(\begin{array}{ccc|c} -1 & -1 & -1 & 0 \\ 0 & 0 & 0 & 0 \\ 0 & 0 & 0 & 0 \end{array} \right)$$

$$\Rightarrow x + y + z = 0$$

$$\Rightarrow x = -y - z$$

$$x = \begin{pmatrix} x \\ y \\ z \end{pmatrix} = \begin{pmatrix} -y - z \\ y \\ z \end{pmatrix} = -y \begin{pmatrix} 1 \\ -1 \\ 0 \end{pmatrix} - z \begin{pmatrix} 1 \\ 0 \\ -1 \end{pmatrix}$$

for all real y and z. Therefore the corresponding eigenvectors are

$$\begin{pmatrix} x \\ y \\ z \end{pmatrix} = \begin{pmatrix} 1 \\ -1 \\ 0 \end{pmatrix} \ and \ \begin{pmatrix} 1 \\ 0 \\ -1 \end{pmatrix}$$

For $\lambda = 2$

$$\begin{pmatrix} 2 & -1 & -1 \\ -1 & 2 & -1 \\ -1 & -1 & 2 \end{pmatrix} \begin{pmatrix} x \\ y \\ z \end{pmatrix} = \begin{pmatrix} 0 \\ 0 \\ 0 \end{pmatrix}$$

$$\left(\begin{array}{ccc|c} 2 & -1 & -1 & 0 \\ -1 & 2 & -1 & 0 \\ -1 & -1 & 2 & 0 \end{array} \right) (R_3 \sim R_1) \left(\begin{array}{ccc|c} -1 & -1 & 2 & 0 \\ -1 & 2 & -1 & 0 \\ 2 & -1 & -1 & 0 \end{array} \right) (R_3 + (R_1 + R_2))$$

$$\left(\begin{array}{ccc|c} -1 & -1 & 2 & 0 \\ -1 & 2 & -1 & 0 \\ 0 & 0 & 0 & 0 \end{array} \right) (R_2 - R_1) \left(\begin{array}{ccc|c} -1 & -1 & 2 & 0 \\ 0 & 3 & -3 & 0 \\ 0 & 0 & 0 & 0 \end{array} \right)$$

$$\Rightarrow y = z \ , \ 2z = x + y$$

$$\Rightarrow x = y = z$$

$$\Rightarrow x = \begin{pmatrix} x \\ y \\ z \end{pmatrix} = \begin{pmatrix} 1 \\ 1 \\ 1 \end{pmatrix}$$

The eigenvector $\begin{pmatrix} 1 \\ 1 \\ 1 \end{pmatrix}$ must be orthogonal to both $\begin{pmatrix} 1 \\ -1 \\ 0 \end{pmatrix} \ and \ \begin{pmatrix} 1 \\ 0 \\ -1 \end{pmatrix}$

We need to apply the Gram-Schmidt process to convert the set

$G = \{(1, -1, 0), \ (1, 0, -1)\}$ into an orthonormal set.

$$\begin{pmatrix} 1 \\ -1 \\ 0 \end{pmatrix}_N = \begin{pmatrix} \dfrac{1}{\sqrt{2}} \\ \dfrac{-1}{\sqrt{2}} \\ 0 \end{pmatrix}$$

$$\begin{pmatrix} 1 \\ 0 \\ -1 \end{pmatrix} \perp \begin{pmatrix} \frac{1}{\sqrt{2}} \\ \frac{-1}{\sqrt{2}} \\ 0 \end{pmatrix} = \begin{pmatrix} 1 \\ 0 \\ -1 \end{pmatrix} - \left\langle (1,0,-1) \left(\frac{1}{\sqrt{2}}, \frac{-1}{\sqrt{2}}, 0 \right) \right\rangle \begin{pmatrix} \frac{1}{\sqrt{2}} \\ \frac{-1}{\sqrt{2}} \\ 0 \end{pmatrix}$$

$$\Rightarrow \begin{pmatrix} 1 \\ 0 \\ -1 \end{pmatrix} \perp \begin{pmatrix} \frac{1}{\sqrt{2}} \\ \frac{-1}{\sqrt{2}} \\ 0 \end{pmatrix} = \begin{pmatrix} 1 \\ 0 \\ -1 \end{pmatrix} - \frac{1}{\sqrt{2}} \begin{pmatrix} \frac{1}{\sqrt{2}} \\ \frac{-1}{\sqrt{2}} \\ 0 \end{pmatrix}$$

$$\Rightarrow \begin{pmatrix} 1 \\ 0 \\ -1 \end{pmatrix} \perp \begin{pmatrix} \frac{1}{\sqrt{2}} \\ \frac{-1}{\sqrt{2}} \\ 0 \end{pmatrix} = \begin{pmatrix} 1 \\ 0 \\ -1 \end{pmatrix} - \begin{pmatrix} \frac{1}{2} \\ \frac{-1}{2} \\ 0 \end{pmatrix}$$

$$= \begin{pmatrix} \frac{1}{2} \\ \frac{-1}{2} \\ 1 \end{pmatrix}$$

$$\begin{pmatrix} \frac{1}{2} \\ \frac{-1}{2} \\ 1 \end{pmatrix}_N = \begin{pmatrix} \frac{1}{\sqrt{6}} \\ \frac{1}{\sqrt{6}} \\ \frac{-2}{\sqrt{6}} \end{pmatrix}$$

Thus the set $G = \left\{ \left(\frac{1}{\sqrt{2}}, \frac{-1}{\sqrt{2}}, 0 \right), \left(\frac{1}{\sqrt{6}}, \frac{1}{\sqrt{6}}, \frac{-2}{\sqrt{6}} \right) \right\}$ is orthonormal

We need also to normalize $\begin{pmatrix} 1 \\ 1 \\ 1 \end{pmatrix}$.

$$\begin{pmatrix} 1 \\ 1 \\ 1 \end{pmatrix}_N = \begin{pmatrix} \frac{1}{\sqrt{3}} \\ \frac{1}{\sqrt{3}} \\ \frac{1}{\sqrt{3}} \end{pmatrix}$$

$$\Rightarrow P = \begin{pmatrix} \frac{1}{\sqrt{3}} & \frac{1}{\sqrt{2}} & \frac{1}{\sqrt{6}} \\ \frac{1}{\sqrt{3}} & \frac{-1}{\sqrt{2}} & \frac{1}{\sqrt{6}} \\ \frac{1}{\sqrt{3}} & 0 & \frac{-2}{\sqrt{3}} \end{pmatrix} ; \ |P| = 1$$

$$D = \begin{pmatrix} 2 & 0 & 0 \\ 0 & -1 & 0 \\ 0 & 0 & -1 \end{pmatrix}$$

$$\Rightarrow D = P^{-1}AP$$

$$X^T A X + N X + 9 = 0$$

becomes

$$X'^T D X' + N P X' + 9 = 0$$

$$\Rightarrow (x', y', z') \begin{pmatrix} 2 & 0 & 0 \\ 0 & -1 & 0 \\ 0 & 0 & -1 \end{pmatrix} \begin{pmatrix} x' \\ y' \\ z' \end{pmatrix} + (-6, -6, 4) \begin{pmatrix} \frac{1}{\sqrt{3}} & \frac{1}{\sqrt{2}} & \frac{1}{\sqrt{6}} \\ \frac{1}{\sqrt{3}} & \frac{-1}{\sqrt{2}} & \frac{1}{\sqrt{6}} \\ \frac{1}{\sqrt{3}} & 0 & \frac{-2}{\sqrt{3}} \end{pmatrix} \begin{pmatrix} x' \\ y' \\ z' \end{pmatrix} + 9 = 0$$

$$\Rightarrow 2x'^2 - y'^2 - z'^2 + (-6, -6, 4) \begin{pmatrix} \frac{x'}{\sqrt{3}} + \frac{y'}{\sqrt{2}} + \frac{z'}{\sqrt{6}} \\ \frac{x'}{\sqrt{3}} - \frac{y'}{\sqrt{2}} + \frac{z'}{\sqrt{6}} \\ \frac{x'}{\sqrt{3}} - \frac{2z'}{\sqrt{6}} \end{pmatrix} + 9 = 0$$

$$\Rightarrow 2x'^2 - y'^2 - z'^2 - 6\left(\frac{x'}{\sqrt{3}} + \frac{y'}{\sqrt{2}} + \frac{z'}{\sqrt{6}}\right) - 6\left(\frac{x'}{\sqrt{3}} - \frac{y'}{\sqrt{2}} + \frac{z'}{\sqrt{6}}\right) + 4\left(\frac{x'}{\sqrt{3}} - \frac{2z'}{\sqrt{6}}\right) + 9 = 0$$

$$\Rightarrow 2x'^2 - y'^2 - z'^2 - \frac{8x'}{\sqrt{3}} = \frac{20z'}{\sqrt{6}} + 9 = 0$$

On completing the square, we have

$$2\left(\left(x' - \frac{2}{\sqrt{3}}\right)^2 - \left(\frac{2}{\sqrt{3}}\right)^2\right) - y'^2 - \left(\left(z' + \frac{10}{\sqrt{6}}\right)^2 - \left(\frac{10}{\sqrt{6}}\right)^2\right) + 9 = 0$$

$$= 2\left(x' - \frac{2}{\sqrt{3}}\right)^2 - \frac{8}{3} - y'^2 - \left(z' + \frac{10}{\sqrt{6}}\right)^2 + \frac{100}{6} + 9 = 0$$

$$\Rightarrow -\frac{\left(x'-\frac{2}{\sqrt{3}}\right)^2}{23/2} + \frac{y'^2}{23} + \frac{\left(z'+\frac{10}{\sqrt{6}}\right)^2}{23} = 1$$

This an elliptic hyperboloid of one sheet oriented along the x''- axis

Where $x'' = \left(x' - \frac{2}{\sqrt{3}}\right)$.

Example 5.19

Name and sketch the graph of the quadratic equation

$x^2 - 4xz + y^2 + 4z^2 - 3x + 2y - 4z = 5$

Solution

In matrix form, the equation is written as

$$X^T A X + N X = 5$$

Where

$$A = \begin{pmatrix} 1 & 0 & -2 \\ 0 & 1 & 0 \\ -2 & 0 & 4 \end{pmatrix}, N = (-3, 2, -4)$$

Let λ be an eigenvalue of A, then

$$|\lambda I - A| = 0$$

$$\begin{vmatrix} \lambda - 1 & 0 & 2 \\ 0 & \lambda - 1 & 0 \\ 2 & 0 & \lambda - 4 \end{vmatrix} = 0$$

Expanding along the second row or column, we have

$$(\lambda - 1)\big((\lambda - 1)(\lambda - 4) - 4\big) = 0$$

$$\lambda(\lambda - 1)(\lambda - 5) = 0$$

$$\Rightarrow \lambda = 0, 1, 5$$

For $\lambda = 0$, we have

$$\begin{pmatrix} -1 & 0 & 2 \\ 0 & -1 & 0 \\ 2 & 0 & -4 \end{pmatrix}\begin{pmatrix} x \\ y \\ z \end{pmatrix} = \begin{pmatrix} 0 \\ 0 \\ 0 \end{pmatrix}$$

$$\Rightarrow y = 0, x = 2z$$

$$\Rightarrow x = \begin{pmatrix} x \\ y \\ z \end{pmatrix} = \begin{pmatrix} 2z \\ 0 \\ z \end{pmatrix} = z\begin{pmatrix} 2 \\ 0 \\ 1 \end{pmatrix} \text{ for all real } z$$

For $\lambda = 1$

$$\begin{pmatrix} 0 & 0 & 2 \\ 0 & 0 & 0 \\ 2 & 0 & -3 \end{pmatrix}\begin{pmatrix} x \\ y \\ z \end{pmatrix} = \begin{pmatrix} 0 \\ 0 \\ 0 \end{pmatrix}$$

$$\Rightarrow y = s, z = 0, x = 0 \text{ for a all real } s$$

$$x = \begin{pmatrix} x \\ y \\ z \end{pmatrix} = \begin{pmatrix} 0 \\ 1 \\ 0 \end{pmatrix}$$

For $\lambda = 5$

$$\begin{pmatrix} 4 & 0 & 2 \\ 0 & 4 & 0 \\ 2 & 0 & 1 \end{pmatrix}\begin{pmatrix} x \\ y \\ z \end{pmatrix} = \begin{pmatrix} 0 \\ 0 \\ 0 \end{pmatrix}$$

$$\Rightarrow y = 0, 2x = -z$$

$$x = \begin{pmatrix} x \\ y \\ z \end{pmatrix} = \begin{pmatrix} x \\ 0 \\ -2x \end{pmatrix} = x\begin{pmatrix} 1 \\ 0 \\ -2 \end{pmatrix} \text{ for all real } x$$

$$\Rightarrow x = \begin{pmatrix} 1 \\ 0 \\ -2 \end{pmatrix}$$

Since the three eigenvectors corresponds to distinct eigenvalues, they must be mutually orthogonal because A is symmetric. Therefore, to assemble a matrix P that orthogonally diagonalizes A, wee need only to normalize each of the eigenvectors and arrange them in columns. Since we want P to represent a rotation, then we must ensure that $|P| = 1$. This means that we arrange our normalized eigenvectors such that $|P| = 1$.

$$\Rightarrow \begin{pmatrix} 2 \\ 0 \\ 1 \end{pmatrix}_N = \begin{pmatrix} \dfrac{2}{\sqrt{5}} \\ 0 \\ \dfrac{1}{\sqrt{5}} \end{pmatrix}$$

$$\begin{pmatrix} 0 \\ 1 \\ 0 \end{pmatrix}_N = \begin{pmatrix} 0 \\ 1 \\ 0 \end{pmatrix} ; \text{ it is already in normal form}$$

$$\begin{pmatrix} 1 \\ 0 \\ -2 \end{pmatrix} = \begin{pmatrix} \dfrac{1}{\sqrt{5}} \\ 0 \\ \dfrac{-2}{\sqrt{5}} \end{pmatrix}$$

$$P = \begin{pmatrix} \dfrac{1}{\sqrt{5}} & 0 & \dfrac{2}{\sqrt{5}} \\ 0 & 1 & 0 \\ \dfrac{-2}{\sqrt{5}} & 0 & \dfrac{1}{\sqrt{5}} \end{pmatrix} ; |P| = 1$$

$$D = P^{-1}AP$$

$$\Rightarrow \quad D = \begin{pmatrix} 5 & 0 & 0 \\ 0 & 1 & 0 \\ 0 & 0 & 0 \end{pmatrix}$$

$$X^T A X + N X = 5$$

becomes

$$X'^T D X' + N P X' = 5$$

$$\Rightarrow 5x'^2 + y'^2 + (-3, 2 - 4) \begin{pmatrix} \dfrac{1}{\sqrt{5}} & 0 & \dfrac{2}{\sqrt{5}} \\ 0 & 1 & 0 \\ \dfrac{-2}{\sqrt{5}} & 0 & \dfrac{1}{\sqrt{5}} \end{pmatrix} \begin{pmatrix} x' \\ y' \\ z' \end{pmatrix} = 5\,5$$

$$\Rightarrow \quad 5x'^2 + y'^2 - 3 \begin{pmatrix} \dfrac{x'}{\sqrt{5}} + \dfrac{2z'}{\sqrt{5}} \\ y' \\ \dfrac{-2x'}{\sqrt{5}} + \dfrac{z'}{\sqrt{5}} \end{pmatrix} = 5$$

$$\Rightarrow \quad 5x'^2 + y'^2 - 3\left(\frac{x'}{\sqrt{5}} + \frac{2z'}{\sqrt{5}}\right) + 2y' - 4\left(\frac{-2x'}{\sqrt{5}} + \frac{z'}{\sqrt{5}}\right) = 5$$

$$\Rightarrow 5x'^2 + \sqrt{5}x' + y'^2 + 2y' - 2\sqrt{5}z' = 5$$

$$\Rightarrow 5\left(\left(x' + \frac{1}{2\sqrt{5}}\right)^2 - \left(\frac{1}{2\sqrt{5}}\right)^2\right) + (y' + 1)^2 - 1^2 - 2\sqrt{5}z' = 5$$

$$\Rightarrow 5\left(x' + \frac{1}{2\sqrt{5}}\right)^2 + (y' + 1)^2 - 2\sqrt{5}z' = 5 + 1 + \frac{1}{4}$$

$$\Rightarrow 5\left(x' + \frac{1}{2\sqrt{5}}\right)^2 + (y' + 1)^2 = 2\sqrt{5}\left(z' + \frac{5\sqrt{5}}{8}\right)$$

$$\Rightarrow \quad \frac{\left(x' + \frac{1}{2\sqrt{5}}\right)^2}{2/\sqrt{5}} + \frac{(y' + 1)^2}{2\sqrt{5}} = \left(z' + \frac{5\sqrt{5}}{8}\right)$$

This, and thus the original equation, is an elliptic paraboloid oriented along the $z'' - $ axis. Where

$$z'' = \left(z' + \frac{5\sqrt{5}}{8}\right)$$

Note: To name a quadratic surface, we don't need to consider the "NPX'" and "f" terms after effecting the transformation $X = PX'$.

We need only to focus on the "$X'^T D'$" term.

Application to a system of Linear Differential Equations

It is often necessary to model reality in terms of equations involving functions and their derivatives. Such equations are called *differential equations.* One of the simplest differential equations is

$$y' = \frac{dy}{dx} = ay$$

We know already that the differential operator is a linear transformation.

Let the differentail linear transformation be denoted as D.

$$\Rightarrow D(y) = y' = ay$$

This means that a is the eigenvalue corresponding to the eigenfunction y.

$$y' = ay$$

$$\Rightarrow \quad y = Ae^{ax}$$

This implies that the eigenfunction corresponding to the eighenvalue a is e^{ax}. In other words, the eigenspace corresponding to the eigenvalue a is spanned by the eigenfunction e^{ax}. When the initial condition is given, the constant A can be found, hence a specific function in the eigenspace of the eigenvalue a. This specific solution is called *particular solution* of the differential equation, while the solution $y = Ae^{ax}$ is called the *general solution.* Now let us consider a more general system of linear differential equations.

$$
\begin{aligned}
y_1' &= a_{11}y_1 + a_{12}y_2 + \ldots\ldots a_{1n}y_n \\
y_2' &= a_{21}y_1 + a_{22}y_2 + \ldots\ldots a_{2n}y_n \\
&\vdots \qquad\qquad \vdots \qquad\qquad \vdots \\
y_n' &= a_{n1}y_1 + a_{n2}y_2 + \ldots\ldots a_{nn}y_n
\end{aligned}
$$

Where the set $(y_1, y_2 \ldots\ldots\ldots y_n)$ is the solution set.

The above linear system can be written in matrix form as

$$Y' = AY \qquad\qquad\qquad -----P226$$

Where

$$
Y' = \begin{pmatrix} y_1' \\ y_2' \\ \vdots \\ \vdots \\ y_n' \end{pmatrix}, \qquad
A = \begin{pmatrix} a_{11} & a_{12} \ldots\ldots\ldots\ldots & a_{1n} \\ a_{21} & a_{22} \ldots\ldots\ldots\ldots & a_{2n} \\ a_{n1} & a_{n2} \ldots\ldots\ldots\ldots & a_{nn} \end{pmatrix}, \quad
Y = \begin{pmatrix} y_1 \\ y_2 \\ \vdots \\ \vdots \\ y_n \end{pmatrix}
$$

Example 5.20

Let the system of linear differential equations be

$$y_1' = 3y_1$$
$$y_2' = -y_2$$
$$y_3' = 5y_3$$

Write the system in matrix form, solve it, and find the particular solution that satisfies the given initial conditions

$$y_1(0) = 2, \quad y_2(0) = 1, \quad y_3(0) = -5$$

Solution

The matrix form, as generally shown above, is

$$\begin{pmatrix} y_1' \\ y_2' \\ y_3' \end{pmatrix} = \begin{pmatrix} 3 & 0 & 0 \\ 0 & -1 & 0 \\ 0 & 0 & 5 \end{pmatrix} \begin{pmatrix} y_1 \\ y_2 \\ y_3 \end{pmatrix}$$

$$Y' = AY$$

The coefficient matrix A is a diagonal matrix, and so each equation in the system can be solved directly, to wit the system is already *decoupled.* This means that each equation in the system involves only one function – there is no mixing. This kind of system, although a system of equations, can be solved individually just as we did for the system $y' = ay$.

$$\therefore \ y_1' = 3y_1, \quad y_2' = -y_2, \quad y_3' = 5y_3$$
$$\Rightarrow y_1 = A_1 e^{3x}, \ y_2 = A_2 e^{-x}, \ y_3 = A_3 e^{5x}$$

To find the particular solution, we apply the given initial conditions

$$y_1 = A_1 e^{3x} \ ; \ 2 = A_1 e^{3(0)} \quad \Rightarrow A_1 = 2$$
$$y_2 = A_2 e^{-x} \ ; \ 1 = A_2 e^{-x(0)} \Rightarrow A_2 = 1$$
$$y_3 = A_3 e^{5x} \ ; -5 = A_3 e^{5(0)} \quad \Rightarrow A_3 = -5$$

$$\Rightarrow y_1 = 2e^{3x} \ , \ \ y_2 = e^{-x} \ , \ \ y_3 = -5e^{5x}$$

This example is somewhat simple to solve because the original problem is already decoupled and so the coefficient matrix is diagonal. However, when at least one equation of the system involves more than one function, the coefficient matrix cannot be diagonal. To solve such a system, we need to decouple it; that is, we need to replace such a system with an equivalent one whose coefficient matrix is diagonal. This is achieved by diagonalizing the original coefficient matrix, and then defining a change of variables. We illustrate this idea, first with a general principle, followed by specific examples.

Consider, again, the general system

$$Y' = AY$$

In which the matrix A is generally nondiagonal. Let A be similar to a diagonal matrix D such that the invertible matrix P diagonalizes A, then

$$D = P^{-1}AP$$

Let's define a change of variables such that

$$Y = PZ$$
$$\Rightarrow Y' = (PZ)'$$
$$\Rightarrow Y' = PZ'; \text{ since } P \text{ is a constant matrix}$$
$$\Rightarrow PZ' = APZ$$
$$\Rightarrow Z' = P^{-1}APZ$$

$$\Rightarrow \qquad\qquad Z' = DZ \qquad\qquad -----P227$$

Example 5.21

Solve the system of linear differential equations

$$y_1' = y_1 + 4y_2$$
$$y_2' = 2y_1 + 3y_2$$
$$y_1(0) = -1 \ , \ y_2(0) = 5$$

Step 1: Write the equation in matrix form

$$Y' = AY$$

Where

$$Y' = \begin{pmatrix} y_1' \\ y_2' \end{pmatrix}, \qquad A = \begin{pmatrix} 1 & 4 \\ 2 & 3 \end{pmatrix}, \qquad Y = \begin{pmatrix} y_1 \\ y_2 \end{pmatrix}$$

Step 2: Find the matrix P that diagonalizes A.

Let λ be an eigenvalue of A, then

$$|\lambda I - A| = 0$$
$$\Rightarrow \begin{vmatrix} \lambda - 1 & -4 \\ -2 & \lambda - 3 \end{vmatrix} = 0$$
$$\Rightarrow (\lambda - 1)(\lambda - 3) - 8 = 0$$
$$\Rightarrow \lambda = 5, -1$$

For $\lambda = 5$

$$\begin{pmatrix} 4 & -4 \\ -2 & 2 \end{pmatrix} \begin{pmatrix} p_1 \\ p_2 \end{pmatrix} = \begin{pmatrix} 0 \\ 0 \end{pmatrix}$$
$$\Rightarrow p_1 = \begin{pmatrix} 1 \\ 1 \end{pmatrix}$$

For $\lambda = -1$

$$\begin{pmatrix} -2 & -4 \\ -2 & -4 \end{pmatrix} \begin{pmatrix} p_1 \\ p_2 \end{pmatrix} = \begin{pmatrix} 0 \\ 0 \end{pmatrix}$$
$$\Rightarrow p_2 = \begin{pmatrix} 2 \\ -1 \end{pmatrix}$$

$$\Rightarrow \quad P = \begin{pmatrix} 1 & 2 \\ 1 & -1 \end{pmatrix}$$

$$D = \begin{pmatrix} 5 & 0 \\ 0 & -1 \end{pmatrix}$$

$$D = P^{-1}AP$$

Step 3: We make the substitution

$$Y = PZ \text{ into } Y' = AY$$

$$\Rightarrow \quad (PZ)' = APZ$$

$$\Rightarrow \quad PZ' = APZ$$

$$\Rightarrow \quad Z' = P^{-1}APZ$$

$$Z' = DZ$$

$$\Rightarrow \begin{pmatrix} z_1' \\ z_2' \end{pmatrix} = \begin{pmatrix} 5 & 0 \\ 0 & -1 \end{pmatrix} \begin{pmatrix} z_1 \\ z_2 \end{pmatrix}$$

Step 4: Solve the system $Z' = DZ$

$$\Rightarrow z_1' = 5z_1 \ , \ z_2' = -z_2$$

$$\Rightarrow \quad z_1 = A_1 e^{5x} \ , \ z_2 = A_2 e^{-x}$$

Step 5: Determine Y from $Y = PZ$

$$\Rightarrow Y = \begin{pmatrix} 1 & 2 \\ 1 & -1 \end{pmatrix} \begin{pmatrix} A_1 e^{5x} \\ A_2 e^{-x} \end{pmatrix}$$

$$Y = \begin{pmatrix} A_1 e^{5x} + 2A_2 e^{-x} \\ A_1 e^{5x} - A_2 e^{-x} \end{pmatrix} = \begin{pmatrix} y_1 \\ y_2 \end{pmatrix}$$

$$\Rightarrow y_1 = A_1 e^{5x} + 2A_2 e^{-x}$$

$$y_2 = A_1 e^{5x} - A_2 e^{-x}$$

This is the general solution. To find the particular solution, we apply the given initial condition as follows.

$$y_1(0) = -1 \quad , \qquad y_2(0) = 5$$

$$A_1 + 2A_2 = -1$$

$$A_1 - A_2 = 5$$

$$\Rightarrow A_1 = 3 \, , A_2 = -2$$

$$\Rightarrow \quad y_1 = 3e^{5x} - 4e^{-x}$$

$$y_2 = 3e^{5x} + 2e^{-x}$$

Example 5.22

Solve the system of linear differential equations

$$y_1' = -y_1 + 2y_2 + 2y_3$$

$$y_2' = 2y_1 - y_2 + 2y_3$$

$$y_3' = 2y_1 + 2y_2 - y_3$$

$$y_1(0) = y_3(0) = 1 \, , y_2(0) = 0$$

Solution

$$\begin{pmatrix} y_1' \\ y_2' \\ y_3' \end{pmatrix} = \begin{pmatrix} -1 & 2 & 2 \\ 2 & -1 & 2 \\ 2 & 2 & -1 \end{pmatrix} \begin{pmatrix} y_1 \\ y_2 \\ y_3 \end{pmatrix}$$

$$Y' = AY$$

$$\Rightarrow A = \begin{pmatrix} -1 & 2 & 2 \\ 2 & -1 & 2 \\ 2 & 2 & -1 \end{pmatrix}$$

We find P that diagonalizes A. Let λ be an eigenvalue of A, then $|\lambda I - A| = 0$

$$\Rightarrow \begin{vmatrix} \lambda+1 & -2 & -2 \\ -2 & \lambda+1 & -2 \\ -2 & -2 & \lambda+1 \end{vmatrix} = 0$$

Expanding along the first row, we have

$$(\lambda + 1)((\lambda + 1)^2 - 4) + 2(-2(\lambda + 1) - 4) - 2(4 + 2(\lambda + 1)) = 0$$

$$\Rightarrow (\lambda + 1)(\lambda - 1)(\lambda + 3) - 8(\lambda + 3) = 0$$

$$(\lambda + 3)(\lambda + 3)(\lambda - 3) = 0$$

$$\lambda = -3 \text{ repeated}, \lambda = 3$$

For $\lambda = 3$

$$\begin{pmatrix} 4 & -2 & -2 \\ -2 & 4 & -2 \\ -2 & -2 & 4 \end{pmatrix} \begin{pmatrix} p_1 \\ p_2 \\ p_3 \end{pmatrix} = \begin{pmatrix} 0 \\ 0 \\ 0 \end{pmatrix}$$

$$\Rightarrow \begin{pmatrix} 4 & -2 & -2 & 0 \\ -2 & 4 & -2 & 0 \\ -2 & -2 & 4 & 0 \end{pmatrix} (R_1 \sim R_3), (R_3 + (R_1 + R_2)) \begin{pmatrix} -2 & -2 & 4 & 0 \\ -2 & 4 & -2 & 0 \\ 0 & 0 & 0 & 0 \end{pmatrix}$$

$$(R_2 - R_1) \begin{pmatrix} -2 & -2 & 4 & 0 \\ 0 & 6 & -6 & 0 \\ 0 & 0 & 0 & 0 \end{pmatrix}$$

$$\Rightarrow p_2 = p_3, \ 2p_3 = p_1 + p_2, \ p_3 = p_1$$

$$\Rightarrow p_1 = \begin{pmatrix} 1 \\ 1 \\ 1 \end{pmatrix}$$

For $\lambda = -3$

$$\begin{pmatrix} -2 & -2 & -2 \\ -2 & -2 & -2 \\ -2 & -2 & -2 \end{pmatrix} \begin{pmatrix} p_1 \\ p_2 \\ p_3 \end{pmatrix} = \begin{pmatrix} 0 \\ 0 \\ 0 \end{pmatrix}$$

$$\begin{pmatrix} -2 & -2 & -2 \\ -2 & -2 & -2 \\ -2 & -2 & -2 \end{pmatrix} (R_2 - R_1), (R_3 - R_1) \begin{pmatrix} -2 & -2 & -2 & 0 \\ 0 & 0 & 0 & 0 \\ 0 & 0 & 0 & 0 \end{pmatrix}$$

$$\Rightarrow p_1 + p_2 + p_3 = 0$$

$$\Rightarrow p = \begin{pmatrix} -p_1 - p_3 \\ p_2 \\ p_3 \end{pmatrix} = -p_2 \begin{pmatrix} 1 \\ -1 \\ 0 \end{pmatrix} - p_3 \begin{pmatrix} 1 \\ 0 \\ -1 \end{pmatrix}$$

$$\Rightarrow p_2 = \begin{pmatrix} 1 \\ -1 \\ 0 \end{pmatrix}, \ p_3 = \begin{pmatrix} 1 \\ 0 \\ -1 \end{pmatrix}$$

$$\Rightarrow P = \begin{pmatrix} 1 & 1 & 1 \\ 1 & -1 & 0 \\ 1 & 0 & -1 \end{pmatrix}$$

$$D = P^{-1}AP$$

$$\Rightarrow D = \begin{pmatrix} 3 & 0 & 0 \\ 0 & -3 & 0 \\ 0 & 0 & -3 \end{pmatrix}$$

We then make the substitution

$$Y = PZ \text{ into } Y' = AY \text{ to have}$$

$$(PZ)' = APZ$$

$$\Rightarrow PZ' = APZ$$

$$\Rightarrow Z' = P'APZ$$

$$Z' = DZ$$

$$\Rightarrow \begin{pmatrix} z_1' \\ z_2' \\ z_3' \end{pmatrix} = \begin{pmatrix} 3 & 0 & 0 \\ 0 & -3 & 0 \\ 0 & 0 & -3 \end{pmatrix}\begin{pmatrix} z_1 \\ z_2 \\ z_3 \end{pmatrix}$$

$$\Rightarrow z_1' = 3z_1, \; z_2' = -3z_2, \quad z_3' = -3z_3$$

$$\Rightarrow z_1 = A_1 e^{3x}, \; z_2 = A_2 e^{-3x}, \; z_3 = A_3 e^{-3x}$$

$$\begin{pmatrix} z_1 \\ z_2 \\ z_3 \end{pmatrix} = \begin{pmatrix} A_1 e^{3x} \\ A_2 e^{-3x} \\ A_3 e^{-3x} \end{pmatrix}$$

Next we determine Y from $Y = PZ$

$$\Rightarrow Y = \begin{pmatrix} 1 & 1 & 1 \\ 1 & -1 & 0 \\ 1 & 0 & -1 \end{pmatrix}\begin{pmatrix} A_1 e^{3x} \\ A_2 e^{-3x} \\ A_3 e^{-3x} \end{pmatrix}$$

$$\Rightarrow \begin{pmatrix} y_1 \\ y_2 \\ y_3 \end{pmatrix} = \begin{pmatrix} A_1 e^{3x} + A_2 e^{-3x} + A_3 e^{-3x} \\ A_1 e^{3x} - A_2 e^{-3x} \\ A_1 e^{3x} - A_3 e^{-3x} \end{pmatrix}$$

$$\Rightarrow y_1 = A_1 e^{3x} + A_2 e^{-3x} + A_3 e^{-3x}$$

$$y_2 = A_1 e^{3x} - A_2 e^{-3x}$$

$$y_3 = A_1 e^{3x} - A_3 e^{-3x}$$

Applying the initial condition

$$y_1(0) = y_3(0) = 1, y_2(0) = 0 \text{, we have}$$

$$A_1 + A_2 + A_3 = 1$$

$$A_1 - A_2 = 0$$

$$A_1 - A_3 = 1$$

$$\Rightarrow \quad A_1 = \frac{2}{3}, A_2 = \frac{2}{3}, A_3 = \frac{-1}{3}$$

$$\Rightarrow \quad y_1 = \frac{1}{3}\left(2e^{3x} + e^{-3x}\right)$$

$$y_2 = \frac{2}{3}\left(e^{3x} - e^{-3x}\right)$$

$$y_3 = \frac{2}{3}e^{3x} + \frac{1}{3}e^{-3x}$$

Recall that

$$\sinh kx = \frac{e^{kx} - e^{-kx}}{2} \quad ; \quad \Rightarrow \quad y_2 = \frac{4}{3}\sinh 3x$$

$$\Rightarrow Y = \begin{pmatrix} y_1 \\ y_2 \\ y_3 \end{pmatrix} = \begin{pmatrix} \frac{1}{3}(2e^{3x} + e^{-3x}) \\ \frac{2}{3}(e^{3x} - e^{-3x}) \\ \frac{1}{3}(2e^{3x} + e^{-3x}) \end{pmatrix}$$

Exponent of A Matrix

We know that

$$e^x = 1 + x + \frac{x^2}{2!} + \frac{x^3}{3!} + \dots \dots$$

$$\Rightarrow e^x = \sum_{k=0}^{\infty} \frac{x^k}{k!}$$

Now when $x = A$, an $n x n$ matrix, we have

$$e^A = I + A + \frac{A^2}{2!} + \frac{A^3}{3!} + \ldots\ldots$$

$$= \sum_{k=0}^{\infty} I \frac{A^k}{k!}$$

In particular, this series converges for $|\lambda| < 1$, where λ is an eigenvalue of A.

Let A be similar to a diagonal matrix D such that the invertible matrix P diagonalizes matrix A.

$$\Rightarrow D = P^{-1}AP$$

$$\Rightarrow A = PDP^{-1}$$

$$\Rightarrow e^A = I + PDP^{-1} + \frac{(PDP^{-1})^2}{2!} + \frac{(PDP^{-1})^3}{3!} + \ldots$$

Recall that

$$(PDP^{-1})^m = PD^mP^{-1}$$

$$\Rightarrow \quad e^A = I + PDP^{-1} + \frac{PD^2P^{-1}}{2!} + \frac{PD^3P^{-1}}{3!} + \ldots$$

$$\Rightarrow \quad e^A = PIP^{-1} + PDP^{-1} \frac{PD^2P^{-1}}{2!} + \frac{PD^3P^{-1}}{3!} + \ldots$$

$$\Rightarrow \quad e^A = P\left(I + D + \frac{D^2}{2!} + \frac{D^3}{3!} + \ldots\right)P^{-1}$$

$$\Rightarrow \qquad\qquad e^A = Pe^DP^{-1} \qquad\qquad\qquad -----P228$$

Let

$$D = diag\,(\lambda_1, \lambda_2, \ldots \lambda_n)$$

We already know that

$$D^m = diag\,(\lambda_1^m, \lambda_2^m, \ldots \lambda_n^m)$$

$$e^D = I + D + \frac{D^2}{2!} + \frac{D^3}{3!} + \ldots$$

$$\Rightarrow \quad e^D = I + diag(\lambda_1, \lambda_2, \ldots \lambda_n) + \frac{1}{2!}diag(\lambda_1^2, \lambda_2^2, \ldots \lambda_n^2) + \frac{1}{3!}diag\,(\lambda_1^3, \lambda_2^3, \ldots \lambda_n^3) + \cdots$$

$$= diag\left(1 + \lambda_1 + \frac{\lambda_1^2}{2!} + \cdots , 1 + \lambda_2 + \frac{\lambda_2^2}{2!} + \cdots , \ldots 1 + \lambda_n + \frac{\lambda_n^2}{2!} + \ldots\right)$$

$$= diag\left(e^{\lambda_1},\, e^{\lambda_2}, \ldots e^{\lambda_n}\right)$$

$$\Rightarrow \qquad e^D = diag\left(e^{\lambda_1}, e^{\lambda_2} \ldots e^{\lambda_n}\right) \qquad ----- P229$$

Example 5.23

Let $A = \begin{pmatrix} 2 & 3 \\ 7 & -2 \end{pmatrix}$, find e^A.

Solution

Let λ be an eigenvalue of A, then

$$|\lambda I - A| = 0$$

$$\Rightarrow \begin{vmatrix} \lambda - 2 & -3 \\ -7 & \lambda + 2 \end{vmatrix} = 0$$

$$\Rightarrow \lambda^2 - 4 - 21 = 0$$

$$\lambda = 5, \lambda = -5$$

For $\lambda = 5$

$$\begin{pmatrix} 3 & -3 \\ -7 & 7 \end{pmatrix}\begin{pmatrix} p_1 \\ p_2 \end{pmatrix} = \begin{pmatrix} 0 \\ 0 \end{pmatrix}$$

$$\Rightarrow p_1 = \begin{pmatrix} 1 \\ 1 \end{pmatrix}$$

For $\lambda = -5$

$$\begin{pmatrix} -7 & -3 \\ -7 & -3 \end{pmatrix}\begin{pmatrix} p_1 \\ p_2 \end{pmatrix} = \begin{pmatrix} 0 \\ 0 \end{pmatrix}$$

$$\Rightarrow p_2 = \begin{pmatrix} 3 \\ -7 \end{pmatrix}$$

$$\Rightarrow \qquad P = \begin{pmatrix} 1 & 3 \\ 1 & -7 \end{pmatrix}$$

$$D = P^{-1}AP$$

$$\Rightarrow \quad D = \begin{pmatrix} 5 & 0 \\ 0 & -5 \end{pmatrix}$$

$$\Rightarrow \quad A = PDP^{-1}$$

$$e^A = Pe^D P^{-1}$$

$$\Rightarrow e^A = \begin{pmatrix} 1 & 3 \\ 1 & -7 \end{pmatrix} \begin{pmatrix} e^5 & 0 \\ 0 & e^{-5} \end{pmatrix} \begin{pmatrix} 1 & 3 \\ 1 & -7 \end{pmatrix}^{-1}$$

$$= \frac{-1}{10} \begin{pmatrix} 1 & 3 \\ 1 & -7 \end{pmatrix} \begin{pmatrix} e^5 & 0 \\ 0 & e^{-5} \end{pmatrix} \begin{pmatrix} -7 & -3 \\ -1 & 1 \end{pmatrix}$$

$$= \frac{1}{10} \begin{pmatrix} 1 & 3 \\ 1 & -7 \end{pmatrix} \begin{pmatrix} 7e^5 & 3e^5 \\ e^{-5} & -e^{-5} \end{pmatrix}$$

$$e^A = \frac{1}{10} \begin{pmatrix} 7e^5 + 3e^{-5} & 3e^5 - 3e^{-5} \\ 7e^5 - 7e^{-5} & 3e^5 + 7e^{-5} \end{pmatrix}$$

We can use the concept of matrix exponent to solve a system of differential equations.

If

$$y' = ay$$

We know that

$$y = Ae^{ax}.$$

Similarly, if

$$Y' = AY$$

Then

$$Y = e^{Ax}B$$

Where $Y' = \begin{pmatrix} y_1' \\ y_2' \\ \vdots \\ y_n' \end{pmatrix}$, $Y = \begin{pmatrix} y_1 \\ y_2 \\ \vdots \\ y_n \end{pmatrix}$, A is an $n x n$ matrix, and B is an $n x 1$ constant vector.

Example 5.24

Solve the system of differential equations

$$y_1' = 2y_1 + 3y_2$$

$$y_2' = 7y_1 - 2y_2 \; ; \; y_1(0) = 0, y_2(0) = 1$$

Solution

$$\begin{pmatrix} y_1' \\ y_2' \end{pmatrix} = \begin{pmatrix} 2 & 3 \\ 7 & -2 \end{pmatrix} \begin{pmatrix} y_1 \\ y_2 \end{pmatrix}$$

$$\Rightarrow \quad Y' = AY$$

$$\Rightarrow \quad Y = e^{Ax}B$$

$$Ax = \begin{pmatrix} 2x & 3x \\ 7x & -2x \end{pmatrix}$$

Recall that if λ is an eigenvalue of A corresponding to the eigenvector $\boldsymbol{p}$, then Ax, x is a scalar, has an eigenvalue of λx corresponding to the same eigenvector $\boldsymbol{p}$.

If

$$A = PDP^{-1}$$

Then

$$Ax = P(Dx)P^{-1}$$

$$\Rightarrow \quad e^{Ax} = Pe^{Dx}P^{-1} \; ; \; \text{since } Dx \text{ is diagonal.}$$

But

$$e^A = \frac{1}{10} \begin{pmatrix} 7e^5 + 3e^{-5} & 3e^5 - 3e^{-5} \\ 7e^5 - 7e^{-5} & 3e^5 + 7e^{-5} \end{pmatrix}$$

$$\Rightarrow \quad e^{Ax} = \frac{1}{10} \begin{pmatrix} 7e^{5x} + 3e^{-5x} & 3e^{5x} - 3e^{-5x} \\ 7e^{5x} - 7e^{-5x} & 3e^{5x} + 7e^{-5x} \end{pmatrix}$$

$$Y = e^{Ax}B$$

Where

$$B = \begin{pmatrix} b_1 \\ b_2 \end{pmatrix}, \text{ a constant vector.}$$

$$\Rightarrow \quad Y = \frac{1}{10}\begin{pmatrix} 7e^{5x} + 3e^{-5x} & 3e^{5x} - 3e^{-5x} \\ 7e^{5x} - 7e^{-5x} & 3e^{5x} + 7e^{-5x} \end{pmatrix}\begin{pmatrix} b_1 \\ b_2 \end{pmatrix}$$

$$\Rightarrow \begin{pmatrix} y_1 \\ y_2 \end{pmatrix} = \begin{pmatrix} c_1(7e^{5x} + 3e^{-5x}) & + & c_2(3e^{5x} - 3e^{-5x}) \\ c_1(7e^{5x} - 7e^{-5x}) & + & c_2(3e^{5x} + 7e^{-5x}) \end{pmatrix}$$

Where

$$c_1 = \frac{b_1}{10}, c_2 = \frac{b_2}{10}$$

$$y_1 = (7c_1 + 3c_2)e^{5x} + (3c_1 - 3c_2)e^{-5x}$$

$$y_2 = (7c_1 + 3c_2)e^{5x} - (7c_1 - 7c_2)e^{-5x}$$

Let

$$q_1 = 7c_1 + 3c_2 \, , \, q_2 = c_1 - c_2$$

$$\therefore \quad y_1 = q_1 e^{5x} + 3q_2 e^{-5x}$$

$$y_2 = q_1 e^{5x} - 7q_2 e^{-5x}$$

To find the particular solution, we apply the initial condition $y_1(0) = 0, y_2(0)1$, we have

$$q_1 + 3q_2 = 0$$

$$q_1 - 7q_2 = 1$$

$$\Rightarrow q_1 = \frac{3}{10} \, , \, q_2 = \frac{-1}{10}$$

$$\Rightarrow \qquad y_1 = \frac{1}{10}(3e^{5x} - 3e^{-5x})$$

$$\Rightarrow \qquad y_1 = \frac{3}{5}\sinh 5x$$

$$\Rightarrow \qquad y_2 = \frac{1}{10}(3e^{5x} + 7e^{-5x})$$

Higher Order Differential Equations

In this section, we briefly show how to use matrix method to solve higher order ordinary linear constant-coefficient homogenous differential equations. To achieve this, we simply convert the given differential equation into a system of first order differential equations. For instance, let us convert the following second order differential equation.

$$a_2 y''(x) + a_1 y'(x) + a_0 y(x) = 0$$

To convert, let

$$z_1(x) = y(x), z_2(x) = y'(x)$$

$$\Rightarrow z_1'(x) = y'(x) = z_2(x)$$
$$z_2'(x) = y''(x)$$

$$\Rightarrow z_1'(x) = z_2(x)$$
$$z_2'(x) = \frac{-a_1}{a_2} z_2(x) - \frac{a_0}{a_2} z_1(x) \qquad ------(a)$$

The above system of linear differential equation (a) is what we seek. To solve it, we can now apply our established matrix method of solving a system of linear differential equations.

$$\Rightarrow \begin{pmatrix} z_1'(x) \\ z_2'(x) \end{pmatrix} = \begin{pmatrix} 0 & 1 \\ -\frac{a_0}{a_2} & -\frac{a_1}{a_2} \end{pmatrix} \begin{pmatrix} z_1(x) \\ z_2(x) \end{pmatrix}$$

Example 5.25

Using the matrix method for linear system, solve the differential equation
$$y'' + y' - 2y = 0.$$

Solution

Let

$$z_1(x) = y(x) \ , \ z_2(x) = y'(x)$$

$$\Rightarrow \ z_1'(x) = y'(x) = z_2(x)$$

$$z_2'(x) = y''(x) = -y' + 2y$$

$$\Rightarrow \ z_1'(x) = z_2(x)$$

$$z_2'(x) = -z_2(x) + 2z_1(x)$$

$$\Rightarrow \ \begin{pmatrix} z_1'(x) \\ z_2'(x) \end{pmatrix} = \begin{pmatrix} 0 & 1 \\ 2 & -1 \end{pmatrix} \begin{pmatrix} z_1(x) \\ z_2(x) \end{pmatrix}$$

$$\Rightarrow \quad Z'(x) = AZ(x)$$

Where

$$Z'(x) = \begin{pmatrix} z_1'(x) \\ z_2'(x) \end{pmatrix} \ , \quad A = \begin{pmatrix} 0 & 1 \\ 2 & -1 \end{pmatrix} , \quad Z(x) = \begin{pmatrix} z_1(x) \\ z_2(x) \end{pmatrix}$$

Let λ be an eigenvalue of A, then

$$|\lambda I - A| = 0$$

$$\Rightarrow \begin{vmatrix} \lambda & -1 \\ -2 & \lambda + 1 \end{vmatrix} = 0$$

$$\Rightarrow \lambda(\lambda + 1) - 2 = 0$$

$$\Rightarrow \lambda = -2 \, , \ \lambda = 1$$

For $\lambda = -2$

$$\begin{pmatrix} -2 & -1 \\ -2 & -1 \end{pmatrix} \begin{pmatrix} p_1 \\ p_2 \end{pmatrix} = \begin{pmatrix} 0 \\ 0 \end{pmatrix}$$

$$\begin{pmatrix} -2 & -1 & | & 0 \\ -2 & -1 & | & 0 \end{pmatrix} (R_2 - R_1) \begin{pmatrix} -2 & -1 & | & 0 \\ 0 & 0 & | & 0 \end{pmatrix}$$

$$\Rightarrow 2p_1 = -p_2$$

$$\Rightarrow \ p_1 = \begin{pmatrix} 1 \\ -2 \end{pmatrix}$$

For $\lambda = 1$

$$\begin{pmatrix} 1 & -1 \\ -2 & 2 \end{pmatrix}\begin{pmatrix} p_1 \\ p_2 \end{pmatrix} = \begin{pmatrix} 0 \\ 0 \end{pmatrix}$$

$$\Rightarrow \; p_1 = p_2$$

$$\Rightarrow \boldsymbol{p_2} = \begin{pmatrix} 1 \\ 1 \end{pmatrix}$$

$$\Rightarrow P = \begin{pmatrix} 1 & 1 \\ -2 & 1 \end{pmatrix}$$

$$D = P^{-1}AP$$

$$\Rightarrow D = \begin{pmatrix} -2 & 0 \\ 0 & 1 \end{pmatrix}$$

As before, we then make the substitution

$$Z(x) = PQ(x)$$

Since

$$Z'(x) = AZ(x)$$

$$\Rightarrow \; \left(PQ(x)\right)' = APQ(x)$$

$$\Rightarrow \; PQ'(x) = APQ(x)$$

$$\Rightarrow \; Q'(x) = P^{-1}APQ(x)$$

$$\Rightarrow \; Q'(x) = DQ(x)$$

$$\Rightarrow \; \begin{pmatrix} q_1'(x) \\ q_2'(x) \end{pmatrix} = \begin{pmatrix} -2 & 0 \\ 0 & 1 \end{pmatrix}\begin{pmatrix} q_1(x) \\ q_2(x) \end{pmatrix}$$

$$\Rightarrow q_1'(x) = -2q_1(x); \; q_2'(x) = q_2(x)$$

$$\Rightarrow q_1(x) = A_1 e^{-2x} \; ; \; q_2(x) = A_2 e^x,$$

We then determine $Z(x)$ from

$$Z(x) = PQ(x)$$

$$\Rightarrow \; \begin{pmatrix} z_1(x) \\ z_2(x) \end{pmatrix} = \begin{pmatrix} 1 & 1 \\ -2 & 1 \end{pmatrix}\begin{pmatrix} A_1 e^{-2x} \\ A_2 e^x \end{pmatrix}$$

$$\Rightarrow \; \begin{pmatrix} z_1(x) \\ z_2(x) \end{pmatrix} = \begin{pmatrix} A_1 e^{-2x} + A_2 e^x \\ -2A_1 e^{-2x} + A_2 e^x \end{pmatrix}$$

$$\Rightarrow z_1(x) = A_1 e^{-2x} + A_2 e^x$$

$$z_2(x) = -2A_1e^{-2x} + A_2e^x$$

Finally we determine $y(x)$

$$y(x) = z_1(x)$$

$$\Rightarrow \qquad y(x) = A_1e^{-2x} + A_2e^x$$

As another way of using matrix exponentiation to solve a linear system of differential equations, let us solve the same problem above using this approach.

$$Z'(x) = AZ(x)$$

$$\begin{pmatrix} z_1'(x) \\ z_2'(x) \end{pmatrix} = \begin{pmatrix} 0 & 1 \\ 2 & -1 \end{pmatrix} \begin{pmatrix} z_1(x) \\ z_2(x) \end{pmatrix}$$

$$Z'(x) = AZ(x)$$

$$Z(x) = e^{Ax}B$$

Where $B = \begin{pmatrix} b_1 \\ b_2 \end{pmatrix}$, a constant vector

$$D = P^{-1}AP$$

$$\Rightarrow \quad A = PDP^{-1}$$

$$\Rightarrow \quad e^{Ax} = Pe^{Dx}P^{-1}$$

$$e^{Ax} = \frac{1}{3} \begin{pmatrix} 1 & 1 \\ -2 & 1 \end{pmatrix} \begin{pmatrix} e^{-2x} & 0 \\ 0 & e^x \end{pmatrix} \begin{pmatrix} 1 & -1 \\ 2 & 1 \end{pmatrix}$$

$$e^{Ax} = \frac{1}{3} \begin{pmatrix} 1 & 1 \\ -2 & 1 \end{pmatrix} \begin{pmatrix} e^{-2x} & -e^{-2x} \\ 2e^x & e^x \end{pmatrix}$$

$$= \frac{1}{3} \begin{pmatrix} e^{-2x} + 2e^x & e^x - e^{-2x} \\ 2e^x - 2e^{-2x} & 2e^{-2x} + e^x \end{pmatrix}$$

$$Z(x) = e^{Ax}B$$

$$Z(x) = \frac{1}{3} \begin{pmatrix} e^{-2x} + 2e^x & e^x - e^{-2x} \\ 2e^x - 2e^{-2x} & 2e^{-2x} + e^x \end{pmatrix} \begin{pmatrix} b_1 \\ b_2 \end{pmatrix}$$

$$\begin{pmatrix} z_1(x) \\ z_2(x) \end{pmatrix} = \frac{1}{3} \begin{pmatrix} b_1(e^{-2x} + 2e^x) & b_2(e^x - e^{-2x}) \\ b_1(2e^x - 2e^{-2x}) & b_2(2e^{-2x} + e^x) \end{pmatrix}$$

$$\Rightarrow z_1(x) = \left(\frac{b_1 - b_2}{3}\right)e^{-2x} + \left(\frac{2b_1 + b_2}{3}\right)e^x$$

$$\Rightarrow z_1(x) = A_1 e^{-2x} + A_2 e^x$$

$$\Rightarrow y(x) = A_1 e^{-2x} + A_2 e^x$$

Example 5.26

Solve the following differential equation using matrix method

$$y''' - 4y'' + y' + 6y = 0$$

Solution

Let

$$z_1(x) = y(x)$$

$$z_2(x) = y'(x)$$

$$z_3(x) = y''(x)$$

$$\Rightarrow \quad z_1'(x) = y'(x) = z_2(x)$$

$$z_2'(x) = y''(x) = z_3(x)$$

$$z_3'(x) = y'''(x) = 4y'' - y' - 6y$$

$$\Rightarrow z_1'(x) = z_2(x)$$

$$z_2'(x) = z_3(x)$$

$$z_3'(x) = 4z_3(x) - z_2(x) - 6z_1(x)$$

$$\Rightarrow \begin{pmatrix} z_1'(x) \\ z_2'(x) \\ z_3'(x) \end{pmatrix} = \begin{pmatrix} 0 & 1 & 0 \\ 0 & 0 & 1 \\ -6 & -1 & 4 \end{pmatrix} \begin{pmatrix} z_1(x) \\ z_2(x) \\ z_3(x) \end{pmatrix}$$

$$Z = A\,Z(x)$$

Where

$$Z'(x) = \begin{pmatrix} z_1'(x) \\ z_2'(x) \\ z_3'(x) \end{pmatrix}, \quad A = \begin{pmatrix} 0 & 1 & 0 \\ 0 & 0 & 1 \\ -6 & -1 & 4 \end{pmatrix}, \quad Z(x) = \begin{pmatrix} z_1(x) \\ z_2(x) \\ z_3(x) \end{pmatrix}$$

Let λ be an eigenvalue of A, then

$$|\lambda I - A| = 0$$

$$\Rightarrow \begin{vmatrix} \lambda & -1 & 0 \\ 0 & \lambda & -1 \\ 6 & 1 & \lambda - 4 \end{vmatrix} = 0$$

Expanding along the first row,

$$\Rightarrow \lambda\left(\lambda(\lambda - 4) + 1\right) + 6 = 0$$

$$\Rightarrow \lambda(\lambda^2 - 4\lambda + 1) + 6 = 0$$

$$\Rightarrow \lambda^3 - 4\lambda^2 + \lambda + 6 = 0$$

$$\Rightarrow \lambda = 1, 2, 3$$

For $\lambda = 1$

$$\begin{pmatrix} 1 & -1 & 0 \\ 0 & 1 & -1 \\ 6 & 1 & -3 \end{pmatrix} \begin{pmatrix} p_1 \\ p_2 \\ p_3 \end{pmatrix} = \begin{pmatrix} 0 \\ 0 \\ 0 \end{pmatrix}$$

$$\Rightarrow p_2 = p_3 = p_1$$

$$\Rightarrow \boldsymbol{p_1} = \begin{pmatrix} 1 \\ 1 \\ 1 \end{pmatrix}$$

For $\lambda = 2$

$$\begin{pmatrix} 2 & -1 & 0 \\ 0 & 2 & -1 \\ 6 & 1 & -2 \end{pmatrix} \begin{pmatrix} p_1 \\ p_2 \\ p_3 \end{pmatrix} = \begin{pmatrix} 0 \\ 0 \\ 0 \end{pmatrix}$$

$$\Rightarrow 2p_2 = p_3 \; ; \; 2p_1 = p_2$$

$$\boldsymbol{p_2} = \begin{pmatrix} 1 \\ 2 \\ 4 \end{pmatrix}$$

For $\lambda = 3$

$$\begin{pmatrix} 3 & -1 & 0 \\ 0 & 3 & -1 \\ 6 & 1 & -1 \end{pmatrix} \begin{pmatrix} p_1 \\ p_2 \\ p_3 \end{pmatrix} = \begin{pmatrix} 0 \\ 0 \\ 0 \end{pmatrix}$$

$$\Rightarrow 3p_2 = p_3; \; 3p_1 = p_2$$

$$\Rightarrow \boldsymbol{p}_3 = \begin{pmatrix} 1 \\ 3 \\ 9 \end{pmatrix}$$

$$\Rightarrow \quad P = \begin{pmatrix} 1 & 1 & 1 \\ 1 & 2 & 3 \\ 1 & 4 & 9 \end{pmatrix}$$

$$D = P^{-1}AP$$

$$D = \begin{pmatrix} 1 & 0 & 0 \\ 0 & 2 & 0 \\ 0 & 0 & 3 \end{pmatrix}$$

Let

$$Z(x) = PQ(x)$$

$$\Rightarrow \quad PQ'(x) = APQ(x)$$

$$\Rightarrow \quad\quad Q'(x) = P^{-1}APQ(x)$$

$$\Rightarrow Q'(x) = DQ(x)$$

$$\Rightarrow q_1'(x) = q_1(x) \ , \ q_2'(x) = 2q_2(x) \ , \ q_3'(x) = 3q_3(x)$$

$$\Rightarrow q_1(x) = A_1 e^x , \ q_2(x) = A_2 e^{2x} , \ q_3(x) = A_2 e^{3x}$$

$$\Rightarrow Q(x) = \begin{pmatrix} q_1(x) \\ q_2(x) \\ q_3(x) \end{pmatrix} = \begin{pmatrix} A_1 e^x \\ A_2 e^{2x} \\ A_3 e^{3x} \end{pmatrix}$$

$$Z(x) = PQ(x)$$

$$\Rightarrow Z(x) = \begin{pmatrix} z_1(x) \\ z_2(x) \\ z_3(x) \end{pmatrix} = \begin{pmatrix} 1 & 1 & 1 \\ 1 & 2 & 3 \\ 1 & 4 & 9 \end{pmatrix} \begin{pmatrix} A_1 e^x \\ A_2 e^{2x} \\ A_3 e^{3x} \end{pmatrix}$$

$$\Rightarrow \begin{pmatrix} z_1(x) \\ z_2(x) \\ z_3(x) \end{pmatrix} = \begin{pmatrix} A_1 e^x + A_2 e^{2x} + A_3 e^{3x} \\ A_1 e^x + 2A_2 e^{2x} + 3A_3 e^{3x} \\ A_1 e^x + 4A_2 e^{2x} + 9A_3 e^{3x} \end{pmatrix}$$

$$\Rightarrow z_1(x) = A_1 e^x + A_2 e^{2x} + A_3 e^{3x}$$

$$\Rightarrow y(x) = A_1 e^x + A_2 e^{2x} + A_3 e^{3x}$$

It is left to the reader to use the exponentiation method to confirm our result.

In general, if

$$a_n y^{(n)}(x) + a_{n-1} y^{(n-1)}_{(x)} + \ldots\ldots a_0 y(x) = 0$$

To convert to a linear system of first order equations, we let

$$z_1(x) = y(x)$$

$$z_2(x) = y'(x)$$

$$\vdots \qquad \vdots$$

$$z_n(x) = y^{(n-1)}(x)$$

So that our linear system becomes

$$z_1'(x) = z_2(x)$$

$$z_2'(x) = z_3(x)$$

$$\vdots \qquad \vdots$$

$$z_{n-1}'(x) = z_n(x)$$

$$z_n'(x) = \frac{-a_{n-1}}{a_n} z_n - \frac{a_{n-2}}{a_n} z_{n-1} - \cdots \frac{a_0}{a_n} z_1$$

So that our matrix form of the system is

$$\begin{pmatrix} z_1'(x) \\ z_2'(x) \\ \vdots \\ \vdots \\ z_n'(x) \end{pmatrix} = \begin{pmatrix} 0 & 1 & 0\ldots\ldots\ldots & 0 \\ 0 & 0 & 1\ldots\ldots. & 0 \\ \vdots & \vdots & 0 & \vdots \\ \vdots & \vdots & \vdots & 1 \\ \frac{-a_0}{a_n} & \frac{-a_1}{a_n} & \frac{-a_2}{a_n}\ldots\ldots. & \frac{-a_{n-1}}{a_n} \end{pmatrix} \begin{pmatrix} z_1(x) \\ z_2(x) \\ \vdots \\ \vdots \\ z_n(x) \end{pmatrix}$$

So that

$$Z'(x) = AZ(x)$$

Where

$$Z'(x) = \begin{pmatrix} z_1'(x) \\ z_2'(x) \\ \vdots \\ \vdots \\ z_n'(x) \end{pmatrix}, \qquad A = \begin{pmatrix} 0 & 1 & 0\ldots\ldots\ldots & 0 \\ 0 & 0 & 1\ldots\ldots. & 0 \\ \vdots & \vdots & 0 & \vdots \\ \vdots & \vdots & \vdots & 1 \\ \frac{-a_0}{a_n} & \frac{-a_1}{a_n} & \frac{-a_2}{a_n}\ldots\ldots. & \frac{-a_{n-1}}{a_n} \end{pmatrix}, Z(x) = \begin{pmatrix} z_1(x) \\ z_2(x) \\ \vdots \\ \vdots \\ z_n(x) \end{pmatrix}$$

Then we let

$$Z(x) = PQ(x)$$

$$\Rightarrow Q'(x) = DQ(x)$$

Where

$$D = P^{-1}AP$$

P is a nonsingular matrix that diagonalizes A

$$D = \begin{pmatrix} \lambda_1 & 0 \dots\dots\dots & & 0 \\ 0 & \lambda_2 & & \vdots \\ \vdots & \vdots & & \vdots \\ 0 & 0 \dots\dots\dots\dots & & \lambda_n \end{pmatrix}$$

$$\Rightarrow \qquad Q'(x) = diag\left(\lambda_1, \lambda_2, \dots\dots \lambda_n\right)(Q(x))$$

$$\Rightarrow q'_j(x) = q_j(x) \; for \; i \leq j \leq n$$

$$\Rightarrow q_j(x) = A_j e^{\lambda_j x} \; for \; 1 \leq j \leq n$$

$$\Rightarrow Z(x) = PQ(x)$$

$$Z(x) = \sum_{j=1}^{n} \boldsymbol{p}_j A_j \, e^{\lambda_j x} \; where \; \boldsymbol{p}_j \; is \; the \; jth \; eigenvector$$

Finally,

$$y(x) = z_1(x)$$

$$\Rightarrow \qquad y(x) = \text{the first row of } Z(x)$$

Where

$$Z(x) = \sum_{j=1}^{n} \boldsymbol{p}_j A_j \, e^{\lambda_j x} \qquad\qquad ----- P230$$

Linear System of Second Order Ordinary Differential Equations

We have seen how diagonalization can be used to decouple or uncouple a coupled linear system of differential equations. Now we show, in a similar way, how a linear system of second order ordinary differential equations can be decoupled. Consider the system

$$y_1'' = a_1 y_1(x) + b_1 y_2(x)$$
$$y_2'' = a_2 y_1(x) + b_2 y_2(x)$$

In matrix form, the system becomes

$$\begin{pmatrix} y_1'' \\ y_2'' \end{pmatrix} = \begin{pmatrix} a_1 & b_1 \\ a_2 & b_2 \end{pmatrix} \begin{pmatrix} y_1 \\ y_2 \end{pmatrix}$$

$$Y'' = AY$$

Where

$$Y'' = \begin{pmatrix} y_1'' \\ y_2'' \end{pmatrix}, \qquad A = \begin{pmatrix} a_1 & b_1 \\ a_2 & b_x \end{pmatrix}, \qquad Y = \begin{pmatrix} y_1 \\ y_2 \end{pmatrix}$$

Let A be similar to the diagonal matrix D such that the nonsingular matrix matrix P diagonalizes A, then

$$D = P^{-1}AP$$

As before, if we make the substitution

$$Y = PZ \text{ , then}$$
$$(PZ)'' = APZ$$
$$\Rightarrow \qquad Z'' = P^{-1}APZ$$
$$\Rightarrow Z'' = DZ$$

Where $D = diag\ (\lambda_1, \lambda_2)$; $\lambda_1\ and\ \lambda_2$ are the eigenvalues of A.

$$\Rightarrow z_1'' = \lambda_1 z_1 \qquad\qquad ------(a)$$
$$z_2'' = \lambda_2 z_2$$

The above system (a) is easily solved, knowing that the differential equation

$$y'' = ay$$

has the solution

$$y = A_1 e^{\sqrt{a}\,x} + A_2 e^{-\sqrt{a}\,x}$$

If $a < 0$, then the general solution above can be written as

$$y = B_1 \cos \sqrt{a}\,x + B_2 \sin \sqrt{a}\,x$$

If $a > 0$, then the general solution can be written as

$$y = B_1 \cosh \sqrt{a}\,x + B_2 \sinh \sqrt{a}\,x$$

Let us illustrate with an example.

Example 5.27

Solve the system

$$y_1'' = 4y_1 + 3y_2$$

$$y_2'' = 2y_1 + 5y_2$$

Where $y_1(0) = 0,\ y_2(0) = 1,\ y_1'(0) = 4,\ y_2'(0) = 1$

Note that $y_1 = y_1(x),\ y_2 = y_2(x)$.

Solution

The system, in matrix form, is

$$\begin{pmatrix} y_1'' \\ y_2'' \end{pmatrix} = \begin{pmatrix} 4 & 3 \\ 2 & 5 \end{pmatrix} \begin{pmatrix} y_1 \\ y_2 \end{pmatrix}$$

$$Y'' = AY$$

Let λ be an eigenvalue of A, then

$$|\lambda I - A| = 0$$

$$\Rightarrow \begin{vmatrix} \lambda - 4 & -3 \\ -2 & \lambda - 5 \end{vmatrix} = 0$$

$$\Rightarrow (\lambda - 4)(\lambda - 5) - 6 = 0$$

$$\lambda^2 - 9\lambda + 14 = 0$$

$$\lambda = 2, 7$$

For $\lambda = 2$

$$\begin{pmatrix} -2 & -3 \\ -2 & -3 \end{pmatrix} \begin{pmatrix} p_1 \\ p_2 \end{pmatrix} = \begin{pmatrix} 0 \\ 0 \end{pmatrix}$$

$$\Rightarrow 2p_1 = -3p_2$$

$$\Rightarrow p_1 = \begin{pmatrix} 3 \\ -2 \end{pmatrix}$$

For $\lambda = 7$

$$\begin{pmatrix} 3 & -3 \\ -2 & 2 \end{pmatrix} \begin{pmatrix} p_1 \\ p_2 \end{pmatrix} = \begin{pmatrix} 0 \\ 0 \end{pmatrix}$$

$$\Rightarrow p_1 = p_2$$

$$\Rightarrow p_2 = \begin{pmatrix} 1 \\ 1 \end{pmatrix}$$

$$\Rightarrow \quad P = \begin{pmatrix} 3 & 1 \\ -2 & 1 \end{pmatrix}$$

$$D = P^{-1}AP$$

$$D = \begin{pmatrix} 2 & 0 \\ 0 & 7 \end{pmatrix}$$

We make the substitution $Y = PZ$, so that

$$(PZ)'' = APZ$$

$$\Rightarrow PZ'' = APZ$$

$$Z'' = P^{-1}APZ$$

$$\Rightarrow Z'' = DZ$$

$$\Rightarrow \begin{pmatrix} z_1'' \\ z_2'' \end{pmatrix} = \begin{pmatrix} 2 & 0 \\ 0 & 7 \end{pmatrix} \begin{pmatrix} z_1 \\ z_2 \end{pmatrix}$$

$$\Rightarrow \quad z_1'' = 2z_1$$

$$z_2'' = 7z_2$$

$$\Rightarrow z_1 = A_1 cosh\sqrt{2}x + A_2 sinh\sqrt{2}x$$

$$z_2 = B_1 cosh\sqrt{7}x + B_2 sinh\sqrt{7}x$$

$$Y = \begin{pmatrix} 3 & 1 \\ -2 & 1 \end{pmatrix} \begin{pmatrix} z_1 \\ z_2 \end{pmatrix} = \begin{pmatrix} y_1 \\ y_2 \end{pmatrix}$$

$$\Rightarrow y_1 = 3z_1 + z_2 \quad , \quad y_2 = -2z_1 + z_2$$

$$\Rightarrow \quad y_1 = 3A_1 cosh\sqrt{2}\,x + 3A_2 sinh\sqrt{2}\,x + B_1 cosh\sqrt{7}x + B_2 sinh\sqrt{7}x$$

$$\Rightarrow \quad y_2 = -2A_1 cosh\sqrt{2}\,x - 2A_2 sinh\sqrt{2}\,x + B_1 cosh\sqrt{7}x + B_2 sinh\sqrt{7}x$$

We now apply the initial condition $y_1(0) = 0$, $y_2(0) = 1$, we have

$$0 = 3A_1 + B_1 \qquad\qquad \text{recall that:} \qquad cosh\ kx = \frac{e^{kx}+e^{-kx}}{2}$$

$$1 = 2A_1 + B_1 \qquad\qquad\qquad\qquad\qquad sinh\ kx = \frac{e^{kx}-e^{-kx}}{2}$$

$$\Rightarrow cosh\ 0 = 1\,, sinh\ 0 = 0$$

$$\Rightarrow A_1 = \frac{-1}{5}\,, \quad B_1 = \frac{3}{5}$$

$$y_1' = 3\sqrt{2}A_1 sinh\sqrt{2}\,x + 3\sqrt{2}\,cosh\sqrt{2}\,x + B_1\sqrt{7}\ sinh\sqrt{7}x + B_2\sqrt{7}\ cosh\sqrt{7}\,x$$

$$y_2' = -2\sqrt{2}A_1 sinh\sqrt{2}\,x - 2\sqrt{2}A_2\,cosh\sqrt{2}\,x + B_1\sqrt{7}\ sinh\sqrt{7}x + B_2\sqrt{7}\ cosh\sqrt{7}\,x$$

Applying $y_1'(0) = 4$, $y_2'(0) = 1$, we have

$$4 = 3\sqrt{2}A_2 + \sqrt{7}B_2$$

$$1 = -2\sqrt{2}A_2 + \sqrt{7}B_2$$

$$\Rightarrow A_2 = \frac{3}{5\sqrt{2}}\,, \quad B_2 = \frac{11}{5\sqrt{7}}$$

$$\Rightarrow y_1 = \frac{-3}{5}cosh\sqrt{2}\,x + \frac{9}{5\sqrt{2}}sinh\sqrt{2}\,x + \frac{3}{5}cosh\sqrt{7}x + \frac{11}{5\sqrt{7}}sinh\sqrt{7}x$$

$$\Rightarrow y_2 = \frac{2}{5}cosh\sqrt{2}\,x - \frac{6}{5\sqrt{2}}sinh\sqrt{2}\,x + \frac{3}{5}cosh\sqrt{7}\,x + \frac{11}{5\sqrt{7}}sinh\sqrt{7}\,x$$

Example 5.28

Solve the system

$$y_1'' = 2y_1 + 7y_2$$

$$y_2'' = y_1 + 3y_2 + y_3$$

$$y_3'' = 5y_1 + 8y_3$$

Where $y_1(0) = 1$, $y_2(0) = 1$, $y_3(0) = 0$, $y_1'(0) = 0$, $y_2'(0) = 0$, $y_3'(0) = 1$.

Note that $y_1 = y_1(x)$, $y_2 = y_2(x)$, $y_3 = y_3(x)$.

Solution

In matrix form, we have

$$\begin{pmatrix} y_1'' \\ y_2'' \\ y_3'' \end{pmatrix} = \begin{pmatrix} 2 & 7 & 0 \\ 1 & 3 & 1 \\ 5 & 0 & 8 \end{pmatrix} \begin{pmatrix} y_1 \\ y_2 \\ y_3 \end{pmatrix}$$

$$Y'' = AY$$

Let λ be an eigenvalue of A, we have

$$|\lambda I - A| = 0$$

$$\Rightarrow \begin{vmatrix} \lambda - 2 & -7 & 0 \\ -1 & \lambda - 3 & -1 \\ -5 & 0 & \lambda - 8 \end{vmatrix} = 0$$

Expanding along the first row, we have

$$(\lambda - 2)(\lambda - 3)(\lambda - 8) + 7(-(\lambda - 8) - 5) = 0$$

$$\lambda^3 - 13\lambda^2 + 39\lambda - 27 = 0$$

$$(\lambda - 1)(\lambda - 3)(\lambda - 9) = 0$$

$$\lambda = 1, 3, 9$$

For $\lambda = 1$

$$\begin{pmatrix} -1 & -7 & 0 \\ -1 & -2 & -1 \\ -5 & 0 & -7 \end{pmatrix} \begin{pmatrix} p_1 \\ p_2 \\ p_3 \end{pmatrix} = \begin{pmatrix} 0 \\ 0 \\ 0 \end{pmatrix}$$

$$p_1 = -7p_2, \qquad 5p_1 = -7p_3$$

$$\Rightarrow p_1 = \begin{pmatrix} 7 \\ -1 \\ -5 \end{pmatrix}$$

For $\lambda = 3$

$$\begin{pmatrix} 1 & -7 & 0 \\ -1 & 0 & -1 \\ -5 & 0 & -5 \end{pmatrix} \begin{pmatrix} p_1 \\ p_2 \\ p_3 \end{pmatrix} = \begin{pmatrix} 0 \\ 0 \\ 0 \end{pmatrix}$$

$$\Rightarrow p_1 = -p_3; \ p_1 = 7p_2$$

$$\Rightarrow p_2 = \begin{pmatrix} 7 \\ 1 \\ -7 \end{pmatrix}$$

For $\lambda = 9$

$$\begin{pmatrix} 7 & -7 & 0 \\ -1 & 6 & -1 \\ -5 & 0 & 1 \end{pmatrix} \begin{pmatrix} p_1 \\ p_2 \\ p_3 \end{pmatrix} = \begin{pmatrix} 0 \\ 0 \\ 0 \end{pmatrix}$$

$$\Rightarrow p_1 = p_2; \ 5p_1 = p_3$$

$$\Rightarrow p_3 = \begin{pmatrix} 1 \\ 1 \\ 5 \end{pmatrix}$$

$$P = \begin{pmatrix} 7 & -7 & 1 \\ -1 & 1 & 1 \\ -5 & -7 & 5 \end{pmatrix}$$

$$D = P^{-1}AP$$

$$D = \begin{pmatrix} 1 & 0 & 0 \\ 0 & 3 & 0 \\ 0 & 0 & 9 \end{pmatrix}$$

Now let

$$Y = PZ$$

$$\Rightarrow \quad (PZ)'' = APZ$$

$$\Rightarrow \quad Z'' = P^{-1}APZ$$

$$Z'' = DZ$$

$$\begin{pmatrix} z_1'' \\ z_2'' \\ z_3'' \end{pmatrix} = \begin{pmatrix} 1 & 0 & 0 \\ 0 & 3 & 0 \\ 0 & 0 & 9 \end{pmatrix} \begin{pmatrix} z_1 \\ z_2 \\ z_3 \end{pmatrix}$$

$$\Rightarrow z_1'' = z_1 \ , \ z_2'' = 3z_2 \ , \ z_3'' = 9z_3$$

$$\Rightarrow z_1 = A_1 \cosh x + A_2 \sinh x$$

$$z_2 = B_1 \cosh \sqrt{3}x + B_2 \sinh \sqrt{3}x$$

$$z_3 = C_1 \cosh 3x + C_2 \sinh 3x$$

$$Y = PZ$$

$$\Rightarrow \begin{pmatrix} y_1 \\ y_2 \\ y_3 \end{pmatrix} = \begin{pmatrix} 7 & 7 & 1 \\ -1 & 1 & -1 \\ -5 & -7 & 5 \end{pmatrix} \begin{pmatrix} z_1 \\ z_2 \\ z_3 \end{pmatrix}$$

$$\Rightarrow y_1 = 7z_1 + 7z_2 + z_3$$

$$\Rightarrow y_2 = -z_1 + z_2 + z_3$$

$$\Rightarrow y_3 = -5z_1 - 7z_2 + 5z_3$$

$$y_1 = 7A_1 \cosh x + 7A_2 \sinh x + 7B_1 \cosh \sqrt{3}x + 7B_2 \sinh \sqrt{3}x + C_1 \cosh 3x + C_2 \sinh 3x$$

$$y_2 = -A_1 \cosh x - A_2 \sinh x + B_1 \cosh \sqrt{3}x + B_2 \sinh \sqrt{3}x + C_1 \cosh 3x + C_2 \sinh 3x$$

$$y_3 = -5A_1 \cosh x - 5A_2 \sinh x - 7B_1 \cosh \sqrt{3}x - 7B_2 \sinh \sqrt{3}x + 5C_1 \cosh 3x + 5C_2 \sinh 3x$$

Applying the initial condition $y_1(0) = y_2(0) = 1$, $y_3(0) = 0$, we have

$$1 = 7A_1 + 7B_1 + C_1$$

$$1 = -A_1 + B_1 + C_1$$

$$0 = 5A_1 - 7B_1 + 5C_1$$

$$\Rightarrow A_1 = \frac{-15}{48} \ , \ B_1 = \frac{5}{12} \ , \ C_1 = \frac{13}{48}$$

$$y_1' = 7A_1 \sinh x + 7A_2 \cosh x + 7\sqrt{3}B_1 \sinh \sqrt{3}x + 7\sqrt{3}B_2 \cosh \sqrt{3}x + 3C_1 \sinh 3x + 3C_2 \cosh 3x$$

$$y_2' = -A_1 \sinh x - A_2 \cosh x + B_1\sqrt{3} \sinh \sqrt{3}x + B_2\sqrt{3} \cosh \sqrt{3}x + 3C_1 \sinh 3x + 3C_2 \cosh 3x$$

$$y_3' = -5A_1 \sinh x - 5A_2 \cosh x$$

$$- 7\sqrt{3}B_1 \sinh \sqrt{3}x - 7\sqrt{3}B_2 \cosh \sqrt{3}x + 15C_1 \sinh 3x + 15C_2 \cosh 3x$$

Applying the initial condition $y_1'(0) = y_2'(0) = 0$, $y_3'(0) = 1$, we have

$$0 = 7A_2 + 7\sqrt{3}\, B_2 + 3C_2$$

$$0 = -A_2 + \sqrt{3}\, B_2 + 3C_2$$

$$1 = -5A_2 - 7\sqrt{3}\, B_2 + 15C_2$$

$$\Rightarrow A_2 = \frac{1}{16} \ , \ B_2 = \frac{-1}{12\sqrt{3}} \ , \ C_2 = \frac{7}{144}$$

$$\Rightarrow y_1 = \frac{-35}{16}\cosh x + \frac{7}{16}\sinh x + \frac{35}{12}\cosh\sqrt{3}x - \frac{7}{12\sqrt{3}}\sinh\sqrt{3}\,x + \frac{13}{48}\cosh 3x + \frac{7}{144}\sinh 3x$$

$$\Rightarrow y_2 = \frac{5}{16}\cosh x - \frac{1}{16}\sinh x + \frac{5}{12}\cosh\sqrt{3}x - \frac{1}{12\sqrt{3}}\sinh\sqrt{3}\,x + \frac{13}{48}\cosh 3x + \frac{7}{144}\sinh 3x$$

$$\Rightarrow y_3 = \frac{25}{16}\cosh x - \frac{5}{16}\sinh x - \frac{35}{12}\cosh\sqrt{3}x + \frac{7}{12\sqrt{3}}\sinh\sqrt{3}\,x + \frac{65}{48}\cosh 3x + \frac{35}{144}\sinh 3x$$

REFERENCES

Adil Yaqub and Hal G. Moore (1980). *Elementary Linear Algebra with applications*. Addison-Wesley Publishing company, Inc.

Seymour Lipschut (1989). *Linear Algebra*. Tata McGraw-Hill Publishing Company Limited, New Delhi.

www.ingramcontent.com/pod-product-compliance
Lightning Source LLC
Chambersburg PA
CBHW080758180526
45168CB00006B/2256